A Practical Introduction to Impedance Matching

Robert L. Thomas

Senior Engineer Scientist

Radiating Systems Design
Avionics Engineering
Douglas Aircraft Company

A Practical Introduction to Impedance Matching

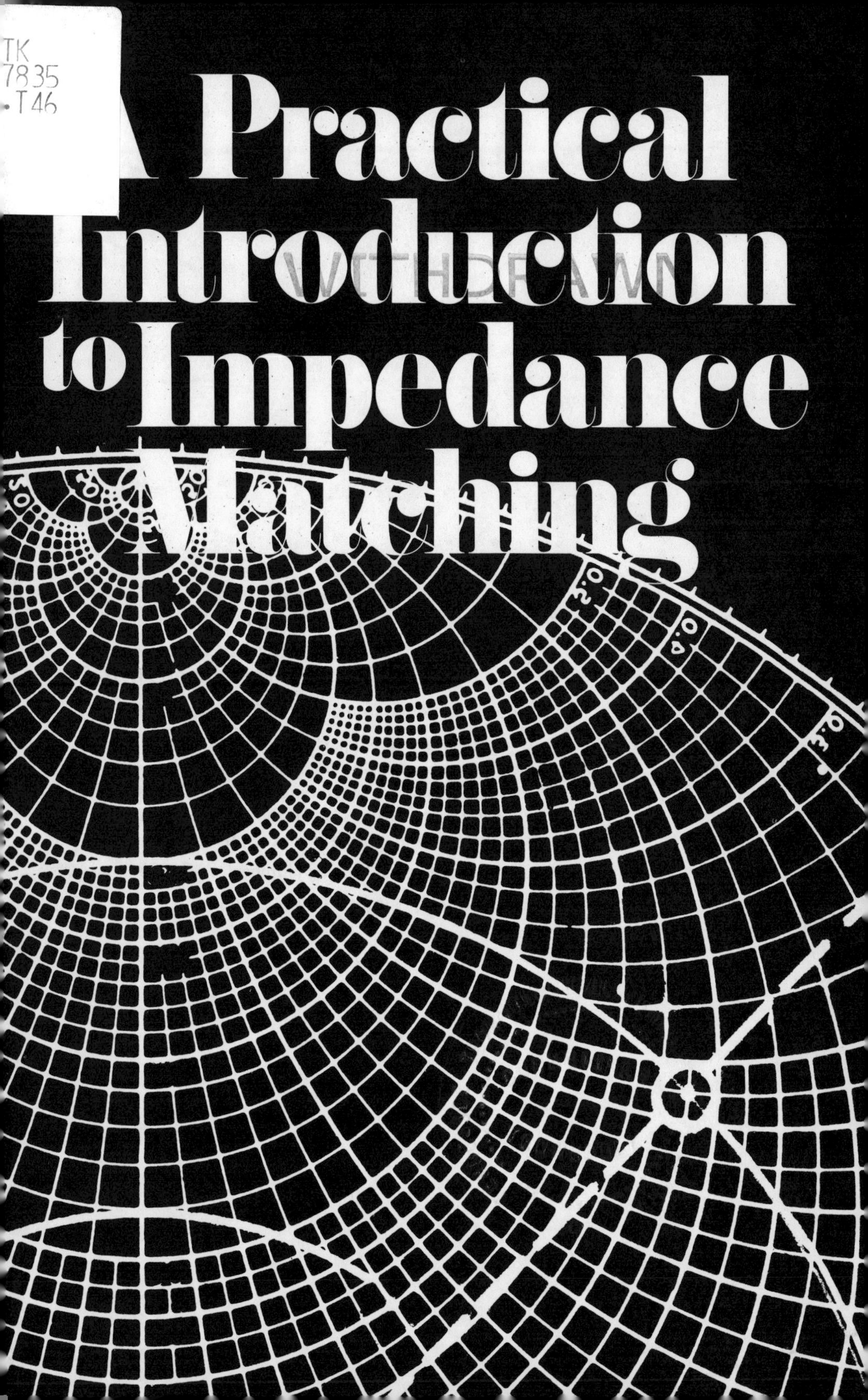

to "Green Eyes"

ARTECH HOUSE, INC.
610 Washington Street
Dedham, Massachusetts 02026

Printed and bound in the United States of America.

Library of Congress Catalog Number: 75-31378

Standard Book Number: 0-89006-050-9

Preface

In this work I have attempted to present a practical introduction to the subject of impedance matching. The material is particularly oriented to those readers who, in their chosen field of endeavor, are involved with either the design, testing, or the specification of performance characteristics for radio frequency devices. No attempt has been made to delve deeply into theoretical aspects nor have complicated mathematical expressions been used. A knowledge of simple algebra is certainly useful but not essential for an understanding of the subject material. Solutions to network problems are accomplished using the well known Smith and Carter Charts. As graphical design aids, these charts are indispensable tools in matching network analysis. This work presents basic impedance relationships, transmission line charts, measurement methods, narrow and broadband matching, and selected practical examples.

I would like to express my appreciation to my leaders, Messrs. William B. Yopp, Melvin F. Gunderson, James R. Burton, and other fellow associates of the Avionics Engineering Section of McDonnell Douglas Corporation, Douglas Aircraft Company, Long Beach, California. Also a special thanks to Jo Lyons who innocently volunteered to type the original manuscript. Working in close association with these fine professional people has provided me with the needed encouragement to complete this technical effort.

ROBERT L. THOMAS

Long Beach, California
December 1973

Table of Contents

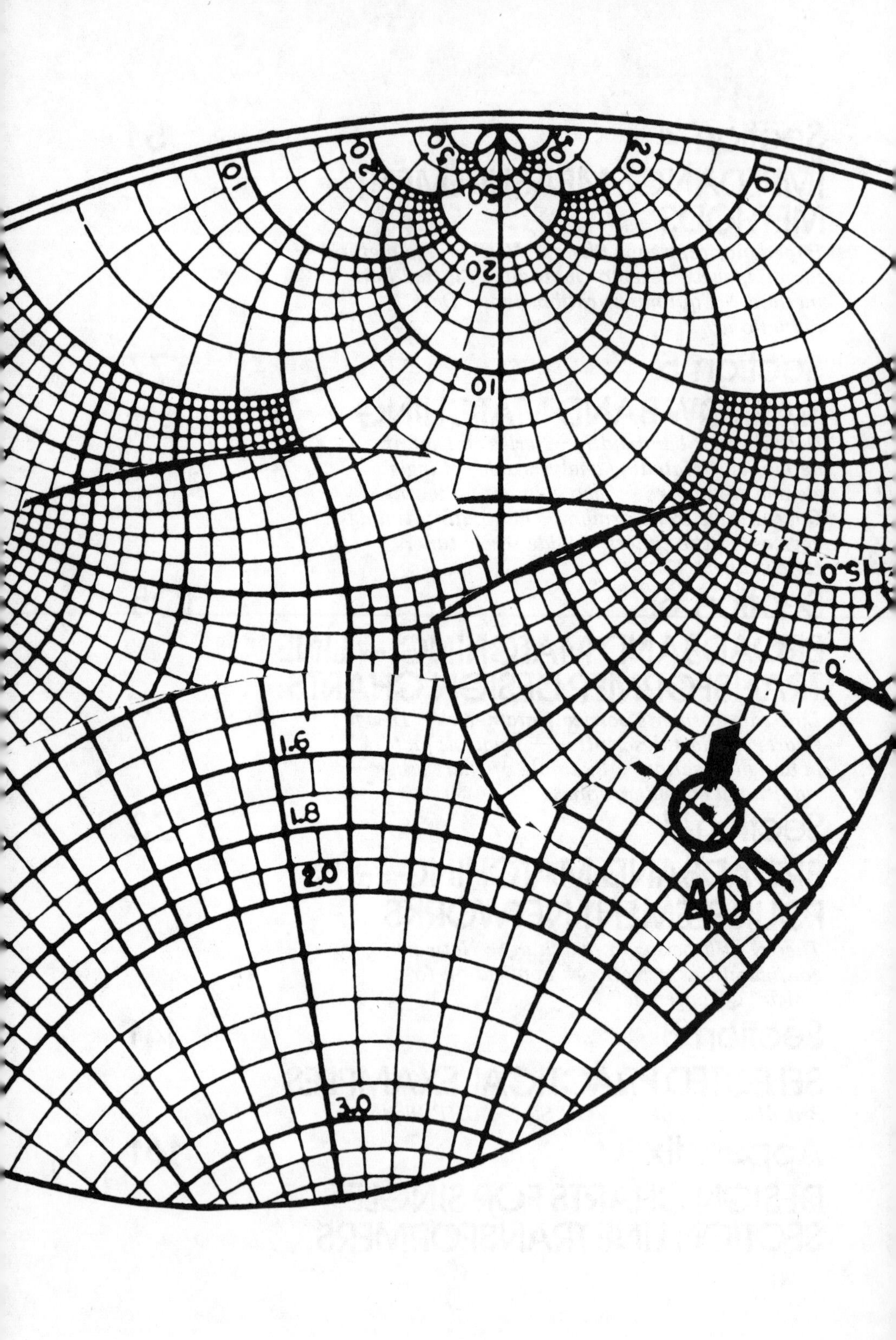
1.6
1.8
2.0
3.0
4.0

1 BASIC RADIO FREQUENCY RELATIONSHIPS

1.1 INTRODUCTION

One of the more interesting aspects of an engineer's task, in the design and development of components for radio frequency (r.f.) systems, is impedance compensation or more commonly termed "matching." Matching simply has the meaning that consideration has been given in the initial design of the r.f. component to assure that the impedance properties of that particular device are compatible with the associated transmission system into which it is to be incorporated. A component that is properly matched will function for efficient transmission of power whereas a badly "mis-matched" component results in an unnecessary loss of energy. In antenna systems, for example, loss of energy as a result of a "mis-matched" antenna may have the meaning that the communication range for that system has been degraded to the extent to be not acceptable.

Impedance matching may for one case be relatively simple to achieve whereas for another quite technically complex. As an illustration, single frequency or narrow band matching requires a minimum of development effort and is generally obtainable with either a single or, at the most, a two-element fixed-tuned network. Broadband matching, on the other hand, is often difficult, particularly in those cases where broad bandwidths and high "Q" structures are involved.

The intent of this work is to serve as a practical introduction to impedance matching and is offered to those who are involved with either the design, testing, or the specification of performance characteristics for r.f. devices. No attempt will be made to delve deeply into theory; only the fundamental aspects of the subject will be considered. This first section is a brief review of the relationships that are considered basic to all radio frequency work.

1.2 WAVELENGTH

The free space wavelength (λo) of a radio frequency wave is given by:

$$\lambda o = \frac{(300)(39.4)}{f} \qquad (1\text{-}1)$$

where: λo = free space wavelength (inches)
f = frequency (megahertz)

The wavelength (λm) in a transmission line filled with a dielectric medium is related to λo as:

$$\lambda m = \frac{\lambda o}{\sqrt{\epsilon}} \qquad (1\text{-}2)$$

Where: λm = wavelength in the medium
ϵ = dielectric constant
For air $\epsilon = 1$

Consider, for example, a teflon ($\epsilon = 2.1$) filled coaxial transmission line whose length is 326 inches and at a frequency of 100 MHz. The wavelength in the line is:

$$\lambda m = \frac{\lambda o}{\sqrt{\epsilon}} = \frac{(300)(39.4)}{(100)(\sqrt{2.1})} = 81.5 \text{ inches}$$

and the electrical length of the line is said to be four wavelengths long.

The product of frequency (f) and (λ) is the velocity of electromagnetic energy and is the speed of light, thus:

$$C = \lambda_0 \cdot f \tag{1-3}$$

and in various units is:

$$\begin{aligned} C &= 3 \times 10^{10} \text{ cm/sec} \\ &= 1.18 \times 10^{10} \text{ in/sec} \\ &= 9.84 \times 10^{8} \text{ ft/sec} \end{aligned}$$

1.3 TRANSMISSION LINES

The primary function of transmission lines is to convey radio frequency energy from one point on the line to another. Some typical transmission lines are shown in Figure 1-1. The wave that travels along the line is an electromagnetic wave formed by electric and magnetic fields that are mutually perpendicular. The electric field is always perpendicular to the surface of a conductor and hence cannot have a component parallel to the surface. The magnetic field, on the other hand, is always parallel to a conductor surface and cannot have a component perpendicular to the surface. An infinite number of waves or modes can exist on a transmission line. Each mode has its own distinctive configuration of electric and magnetic fields that satisfy Maxwell's equations and fit the boundary conditions imposed by the line.

Transmission line modes may be separated into three classes such as:

- Principal Mode — This mode can occur only for lines containing not more than two separated conductors and consists only of electric and magnetic fields that are everywhere transverse to the

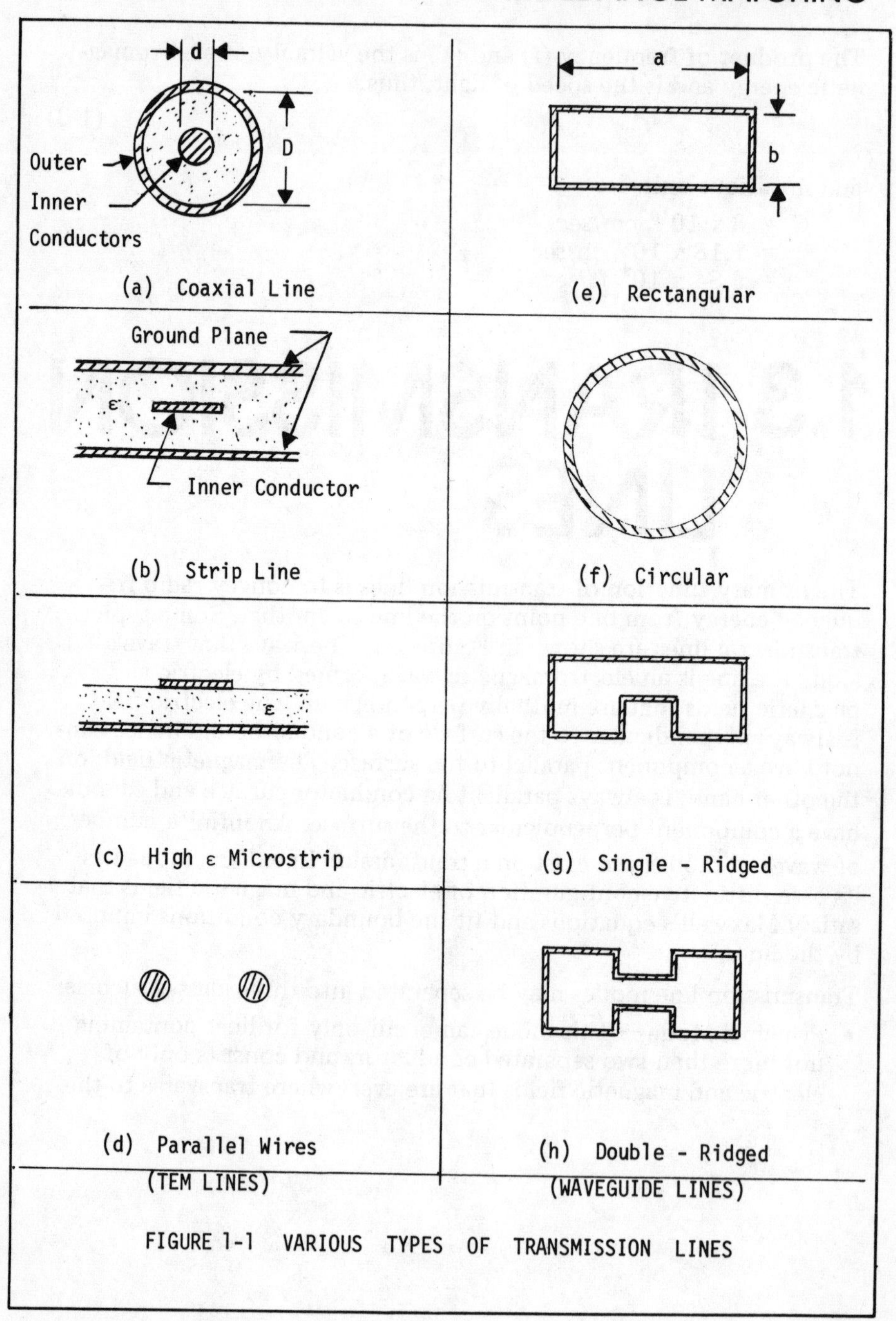

FIGURE 1-1 VARIOUS TYPES OF TRANSMISSION LINES

direction of energy flow. The mode is referred to as TEM signifying a Transverse Electromagnetic Wave. The coaxial line is the most common example of this mode of transmission.

Higher order modes (called waveguide modes) can exist in a coaxial line if, in some manner, the line is excited at a frequency above cut-off for that mode. Higher order modes may be excited below mode cut-off by a discontinuity in the line but such waves will attenuate very rapidly with distance as they travel along the line. Coaxial lines are generally designed to operate only in the principal mode — the approximate limiting frequency, or cut-off frequency for the first higher order TE mode (TE_{11}), given as:

$$f = \frac{7.51 \times 10^3}{\sqrt{\epsilon} \cdot (d + D)}$$

Where: f = limiting frequency (megahertz)
d, D = as defined in Figure 1-1 (inches)

- Transverse Electric (TE) modes of transmission have magnetic field components in the direction of energy flow whereas the electric field is everywhere transverse.
- Transverse Magnetic (TM) modes have electric field components in the direction of energy flow and the magnetic field is everywhere transverse.

TE and TM waveguide modes are termed "higher order" modes and have low frequency cut-off limits below which energy will not be propagated along the line. The higher order modes are further distinguished by two subscripts. The first subscript indicates the number of half-wave variations of the electric field across the wide dimension of the guide. The second subscript describes the number across the narrow dimension of the guide. For example, the dominant mode (lowest frequency waveguide mode) in rectangular waveguide is the $TE_{1,0}$ mode. Figure 1-2 shows the field configurations for the principal (coaxial) and dominant (simplest waveguide) modes of transmission.

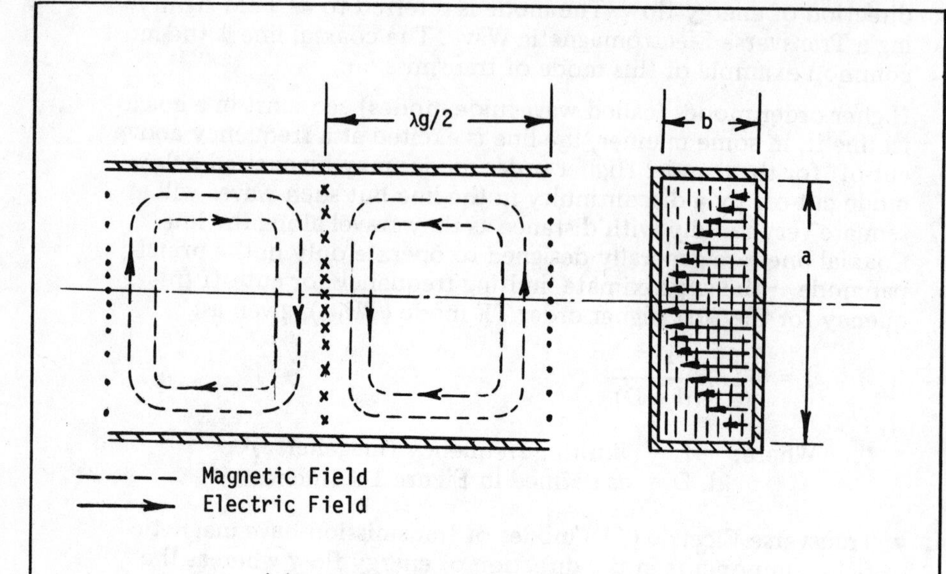

(a) Waveguide Dominant Mode ~ $TE_{1,0}$

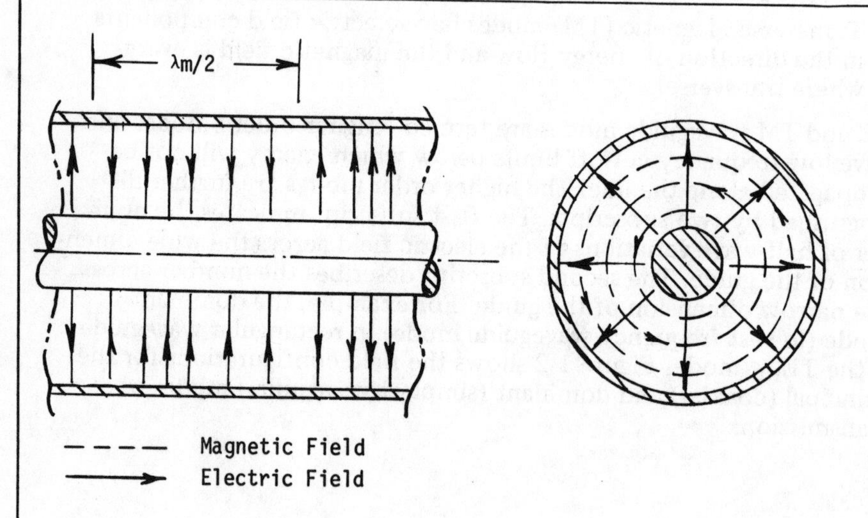

(b) Coaxial Principal Mode ~ TEM

FIGURE 1-2 | FIELD CONFIGURATIONS FOR TEM AND $TE_{1,0}$ MODES

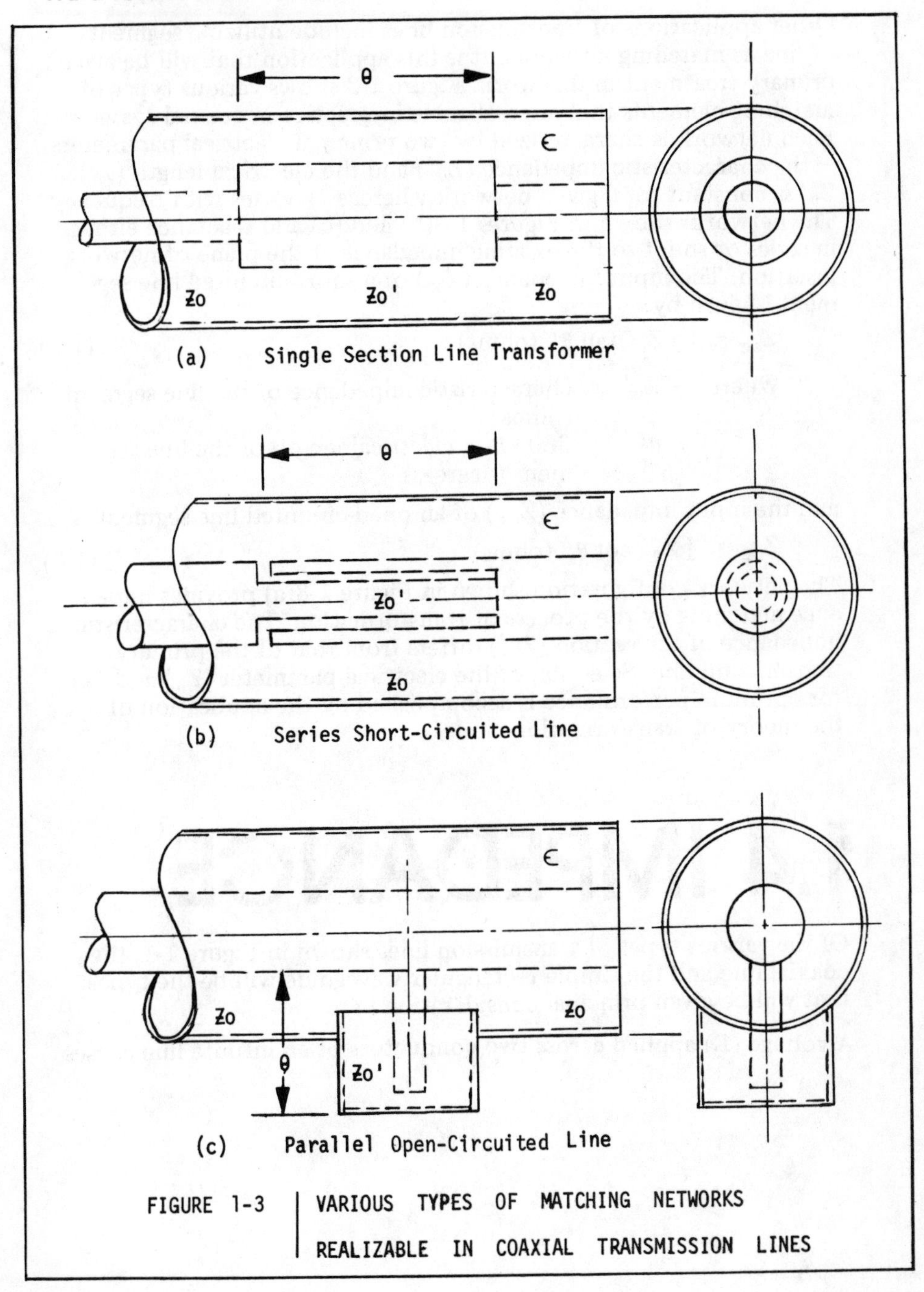

FIGURE 1-3 | VARIOUS TYPES OF MATCHING NETWORKS REALIZABLE IN COAXIAL TRANSMISSION LINES

Other applications of transmission lines include utilizing segments of line as matching networks. It is this application that will be given primary treatment in this work. Figure 1-3 shows various types of matching elements that are realizable in practice in coaxial systems. Each network is characterized by two principal electrical parameters — the characteristic impedance (Z_0') and the electrical length ($\theta \ell$). Z_0' is constant for a given network whereas $\theta \ell$ varies with frequency. The networks shown in Figures 1-3(b) and (c) add reactance either in series or shunt to the existing impedance at the plane of network insertion. The input impedance (Z_{sc}) of a short-circuited line segment is given by:

$$Z_{sc} = +j\, Z_0' \tan \theta \ell \text{ (ohms)} \tag{1-5}$$

Where: Z_0' = characteristic impedance of the line segment (ohms)
$\theta \ell$ = $360\, \ell/\lambda$ = electrical length of the line segment (degrees)

and the input impedance (Z_{oc}) of an open-circuited line segment is:

$$Z_{oc} = -j\, Z_0' \cot \theta \ell \text{ (ohms)} \tag{1-6}$$

The network configuration shown in Figure 1-3(a) provides impedance matching by the process of transformation. The characteristic impedance of the section (Z_0') differs from that of the primary transmission line. Selection of the electrical parameter (Z_0' and $\theta \ell$) for optimum performance is accomplished by the application of the theory of transformation circles.

1.4 IMPEDANCE

Of the various types of transmission lines shown in Figure 1-1, the coaxial line and the simple rectangular waveguide will be the types that will be given principal consideration here.

A voltage (E) applied across two conductors of an infinite line causes a current (I) to flow. The term characteristic (Z_0) impedance (or surge impedance) is derived from this relationship as:

$$Z_0 = \frac{E}{I} = \sqrt{\frac{Z}{Y}} = \sqrt{\frac{R + j\omega L}{G + j\omega C}} \quad \text{(ohms)} \tag{1-7}$$

Where: $Z = R + j\omega L$ = series impedance of a unit section of the infinite line

$Y = G + j\omega C$ = shunt admittance of a unit section of the infinite line

and for a "lossless" line — that is one in which the attenuation may be considered negligible:

$$Z_0 = \sqrt{\frac{L}{C}} \quad \text{(ohms)} \tag{1-8}$$

Where: L = inductance per unit length (henries)
C = capacitance per unit length (farads)

The characteristic impedance of the coaxial line (Figure 1-1) is derived from the dielectric constant (ϵ) and line geometry as:

$$Z_0 = \frac{138}{\sqrt{\epsilon}} \log_{10} \frac{D}{d} \quad \text{(ohms)} \tag{1-9}$$

The characteristic impedance of a segment of transmission line of unknown value may be determined by measurement of two equal segments of the line where one segment is terminated in a short circuit and the other segment terminated in an open circuit. The Z_0 is calculated as:

$$Z_0 = \sqrt{Z_{oc} \cdot Z_{sc}} \quad \text{(ohms)} \tag{1-10}$$

Where: Z_{oc} = open circuit impedance
Z_{sc} = short circuit impedance

The characteristic impedance of a uniform loss less coaxial line is always real and does not differ with frequency.

Waveguide characteristic impedance differs with frequency and for simple rectangular guide, operating in the $TE_{1,0}$ mode, is:

$$Z_0 = \frac{60\pi^2}{\sqrt{\epsilon}} \cdot \frac{b}{a} \cdot \frac{1}{\sqrt{1 - (\frac{fc}{f})^2}} \text{ (ohms)} \qquad (1\text{-}11)$$

Where: a, b = cross-sectional dimensions (see Figure 1-1)
fc = cut-off frequency
f = frequency at which Z_0 is desired

and the cut-off wavelength (λ_c) is:

$$\lambda c = 2a \qquad (1\text{-}12)$$

The guide wavelength (λg) also differs with frequency as:

$$\lambda g = \lambda \cdot \frac{1}{\sqrt{1 - (\lambda / \lambda_c)^2}} \qquad (1\text{-}13)$$

Where: λ = wavelength in the unbounded medium of dielectric constant ϵ.

Figure 1-4 shows various matching devices that are useable in waveguide type structures.

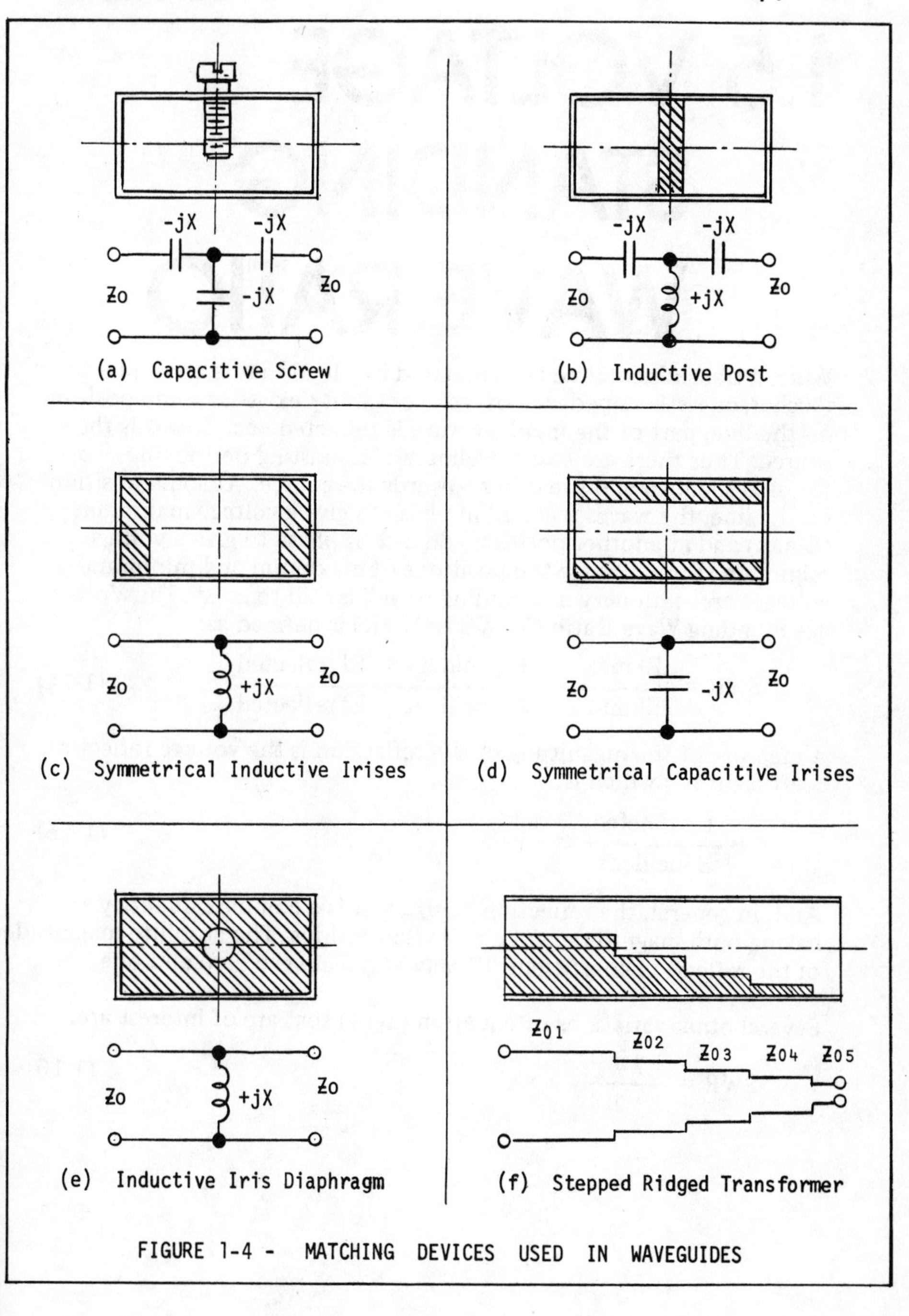

FIGURE 1-4 - MATCHING DEVICES USED IN WAVEGUIDES

1.5 VOLTAGE STANDING WAVE RATIO

When a transmission line is terminated in a load that is not equal to its characteristic impedance or a discontinuity exists at some position on the line, part of the incident wave is reflected back towards the source. Thus there are two traveling waves existing on the line — one towards the load and the other towards the source. At some position on the line, the waves will add in phase to give a voltage maximum (Emax) and at another position add out of phase to give a voltage minimum (Emin). Since the positions of maximum and minimum voltages are stationary a "standing wave" is said to exist. The Voltage Standing Wave Ratio (VSWR or SWR) is defined as:

$$SWR = \frac{|E|\,max}{|E|\,min} = \frac{|E|\,incident + |E|\,reflected}{|E|\,incident - |E|\,reflected} \qquad (1\text{-}14)$$

A measure of the magnitude of the reflection is the voltage reflection coefficient (Γ) which is:

$$\Gamma = \frac{E\,reflected}{E\,incident} = |\Gamma| \angle\phi \qquad (1\text{-}15)$$

And, in general, the reflection coefficient is a complex quantity — having both magnitude and phase relationships. The absolute magnitude of the reflection coefficient |Γ| may vary between zero and one.

Several other variations of equation (1-14) that are of interest are:

$$SWR = \frac{1 + |\Gamma|}{1 - |\Gamma|} \qquad (1\text{-}16)$$

or

$$|\Gamma| = \frac{SWR - 1}{SWR + 1} \qquad (1\text{-}17)$$

The absolute magnitude of the reflection coefficient may be utilized to describe the efficiency of transmission (or percent of energy reflected) in the following relationship:

$$\%\ \text{Power Reflected} = |\Gamma|^2 \cdot 100 \qquad (1\text{-}18)$$

Consider, for example, an antenna system with an input SWR of 5:1, which is typical of VOR communication systems operating in the frequency range of 108 MHz to 118 MHz. Such a system would have a reflection coefficient of 0.67 with approximately 45% of the input power reflected back into the transmission system as wasted energy. Thus the importance of impedance matching, to meet specified performance requirements for SWR, is easily recognized when related to energy loss.

1.6 ATTENUATION

The SWR on a "lossless" transmission line is a constant value. Traveling a half-wavelength ($\lambda/2$) on such a line transcribes a circle on the Smith Chart which is a graphical design aid for the solution of transmission line problems. All practical transmission lines have finite attenuation and hence losses. Attenuation has the meaning that signal traveling along the line is reduced in magnitude as it progresses. The reflection coefficient (and hence SWR) also differ as measured at different points on a line containing attenuation. The reflection coefficient at any position along the line ($|\Gamma_p|$) is related to the load reflection coefficient ($|\Gamma_L|$) as:

$$|\Gamma_p| = |\Gamma_L|\epsilon^{-2\alpha\ell} \qquad (1\text{-}19)$$

or

$$|\Gamma_L| = |\Gamma_p|\epsilon^{+2\alpha\ell} \qquad (1\text{-}20)$$

Where: α = attenuation in nepers/meter
ℓ = length in meters

In practice, published attenuation data for transmission lines is expressed in decibels (db) per specified length. For example, RG-9/U coaxial cable has about 2.1 db per 100 feet at 100 MHz, α in (1-19) and (1-20) is in neper units. The relationship between nepers and db is:

$$1 \text{ neper} = 8.68 \text{ db} \tag{1-21}$$

The term (db) is also used in radio frequency work to relate the ratios of two voltages or two powers as:

$$\text{db} = 20 \log_{10} \frac{E1}{E2} \tag{1-22}$$

and

$$\text{db} = 10 \log_{10} \frac{P1}{P2} \tag{1-23}$$

1.7 TRANSMISSION LINE EQUATIONS

The input impedance (Z_{in}) "looking into" a "lossless" transmission line terminated in a load (Z_L) is given by:

$$Z_{in} = Z_0 \frac{Z_L + j Z_0 \tan \theta \ell}{Z_0 + j Z_L \tan \theta \ell} \tag{1-24}$$

Where: Z_0 = line characteristic impedance (ohms)
Z_L = load impedance (ohms)
$\theta \ell$ = electrical length of line (degrees)

The input impedance is generally a complex quantity. The process of impedance matching requires numerous manipulations of impedance quantities in the derivation of optimum networks. In such operations, application of equation (1-24) is quite tedious and time consuming. Fortunately there are other methods less tedious and more readily applied. Such methods include the utilization of graphical charts (Smith and Carter)[2,3] that not only provide ready solutions to transmission line problems but also "visual" interpretation of parameter variation along a given line. The work herein will treat the solution of network problems by graphical techniques only. Equation (1-24) is given for technical interest.

REFERENCES

1. J.A. Nelson and G. Stavis, "Impedance Matching, Transformers and Baluns," *Very High Frequency Techniques*, Chapter 3, Radio Research Laboratory, McGraw-Hill Book Company, New York and London, 1947.
2. P.H. Smith, "Transmission Line Calculator," *Electronics Magazine*, January 1939.
3. P.S. Carter, "Charts for Transmission Line Measurements and Computations, *R.C.A. Review*, No. 3, p. 355, January 1939.

BIBLIOGRAPHY

1. Theodore Moreno, *Microwave Transmission Design Data*, Dover Publications, Inc., New York, New York, 1958.
2. Gershon J. Wheeler, *Introduction to Microwaves*, Prentice-Hall, Inc., New Jersey, 1963.
3. Theodore S. Saad, *Microwave Engineers Handbook*, Volumes I and II, Artech House, Inc., Dedham, Massachusetts, 1971.
4. M. Slurzberg and W. Osterheld, *Essentials of Radio*, McGraw-Hill Book Company, Inc., New York — Toronto — London, 1948.
5. "Transmission Lines," *Reference Data for Radio Engineers*, Chapter 20, International Telephone and Telegraph Corporation, fourth edition, 1957.

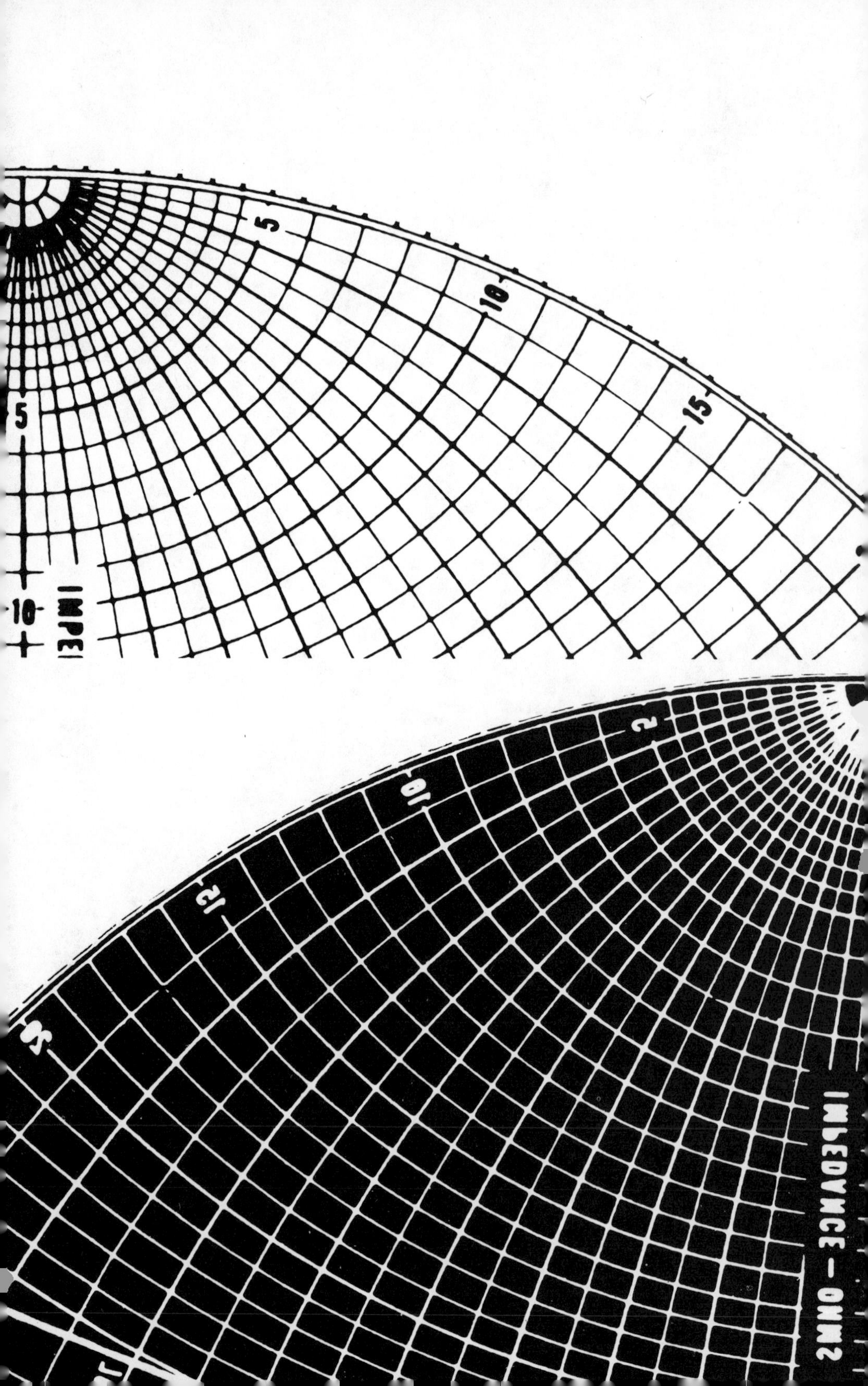

5
10
15

2 TRANSMISSION LINE CHARTS—1

2.1 INTRODUCTION

The basic transmission line equations permit the determination of the relationships between voltages, currents, impedances, reflection coefficient, and other related transmission line parameters and the manner in which these quantities vary along a given line. Equation (1-24) shows the case for the "lossless line." Other cases are given in literature[1]. The mathematical process involved in the application of these equations, however, is such to not readily lend itself to the rapid solution to practical problems that engineers and technicians encounter in their daily work.

Simple transmission line charts are available [2] that reduce the mathematical labor involved in the solution of transmission line problems. In the field of radio frequency measurements and component development, these charts are universally recognized as the proper format for the reporting and recording of impedance data. Perhaps their most useful application, however, is as graphical design aids in matching network analysis. The two more popular transmission line charts are the Smith and Carter charts. Both charts are circular in form. The Smith Chart is a polar plot of reflection coefficient with a coordinate system superimposed that represents the rectangular form of impedances or admittances. The Carter Chart coordinate system represents the polar form.

Sections II and III will be concerned with the above two types of transmission line charts. Together the two sections present a basic description of the charts together with a discussion of the fundamental impedance operations that may be performed. The reader thus obtains the knowledge of a design tool that is most useful in the graphical derivation of solutions to impedance matching network problems.

2.2 SMITH CHART

2.2.1 Coordinate Grid System

The Smith Chart has a coordinate grid system that is arranged in such a manner to permit the plotting, in rectangular form, of all values of impedances (R ± j X) and admittances (G ± j B). The conventional form of the chart, Figure 2-1, is generally used for impedance operations where SWR's of greater than 2:1 are involved. An expanded version of the chart, not shown, enlarges the center portion and has application for operations where SWR's are less than 2.6:1.

The grid system is said to be "normalized" as the prime center of the chart, on the vertical resistance/conductance axis, is given a value of 1.0. This represents the 100% value of the characteristic impedance (Z_0) or admittance (Y_0) of the associated transmission line. The scale of the resistance/conductance axis ranges from zero to infinity and normalized value appearing on this axis represent percentages of the value assigned to the prime center.

Reference to Figure 2-1 shows that there are two families of normalized orthogonal circles — the resistance/conductance circles and the reactance/susceptance circles. The resistance/conductance circles are centered upon the vertical axis and represent the real component of impedance/admittance. The reactance/susceptance circles appear on the right and left sides of the chart and represent the imaginary component of impedance/admittance. Sufficient normalized circles are shown on the Smith Chart to permit a practical interpolation accuracy when a plotted value falls between two given circles.

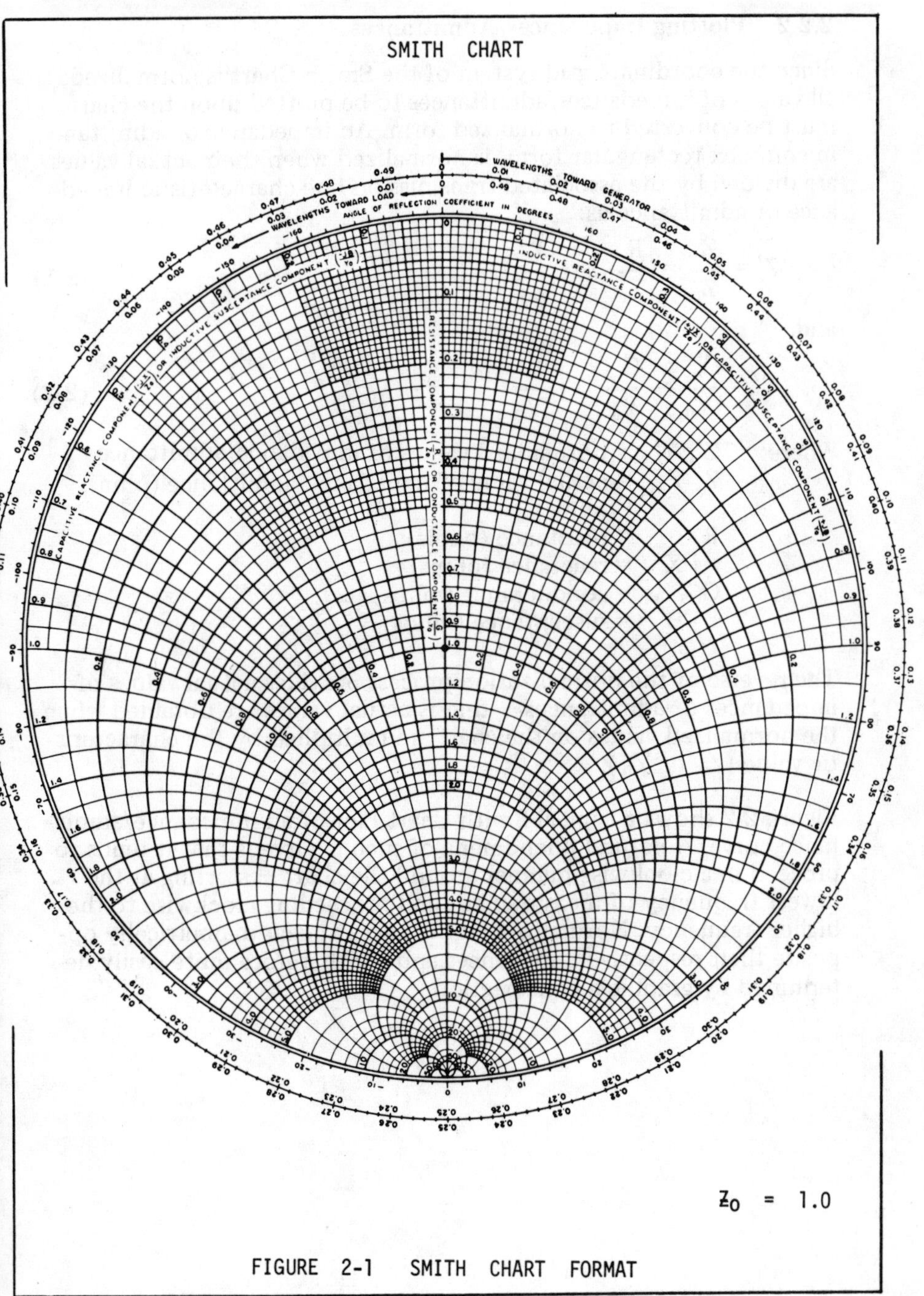

FIGURE 2-1 SMITH CHART FORMAT

2.2.2 Plotting Impedances/Admittances

Since the coordinate grid system of the Smith Chart is normalized, all values of impedances/admittances to be plotted upon the chart must be converted to normalized form. An impedance or admittance, in complex rectangular form, is normalized when their actual values are divided by the associated transmission line characteristic impedance or admittance as:

$$Z' = \frac{Z}{Z_0} = \frac{R}{Z_0} \pm j \frac{X}{Z_0} \tag{2-1}$$

and

$$Y' = \frac{Y}{Y_0} = \frac{G}{Y_0} \pm j \frac{B}{Y_0} \tag{2-2}$$

Where:
- Z', Y' = normalized values of impedance/admittance
- Z, Y = actual value of impedance/admittance (ohms/mhos)
- R = resistance (ohms)
- $\pm jX$ = reactance (ohms)
- G = conductance (mhos)
- $\pm jB$ = susceptance (mhos)

The reverse of the normalization process provides actual values of impedances or admittances — that is actual values are obtained when the normalized values on the chart are multiplied by the characteristic value (Z_0 or Y_0).

Figure 2-2 shows a typical Smith Chart Plot of an impedance/admittance curve. A natural phenomena of such a curve is that it tends to proceed in a clockwise manner around the chart — starting at the lowest frequency of measurement and proceeding clockwise to the higher frequency. Normalized admittances appear diametrically opposite their respective impedances and are therefore quite easily determined by graphical construction.

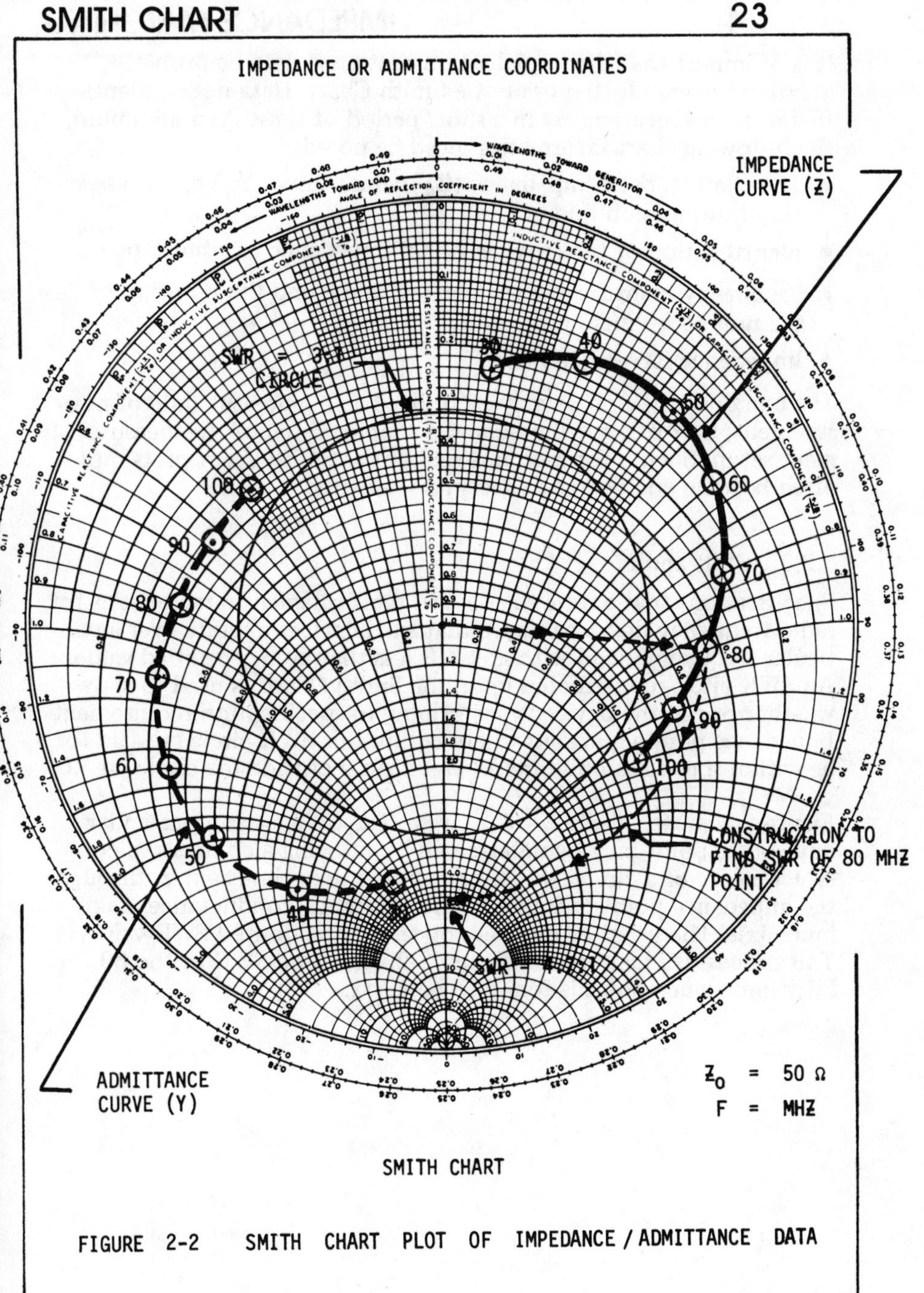

FIGURE 2-2 SMITH CHART PLOT OF IMPEDANCE / ADMITTANCE DATA

It is of import that the impedance/admittance data be properly identified when plotted upon the Smith Chart. Data not so identified becomes meaningless in a short period of time. As a minimum, the following chart information need be noted:

- The characteristic impedance (Z_0)/admittance (Y_0) of the associated transmission line.
- Identification of the curve as either impedance or admittance.
- Frequency identification of plotted values and the notation of the unit of frequency.
- Impedance Reference Plane.

The margins of the chart may be utilized to include the date of measurement, network schematic, and a simple sketch of the measurement setup. Such information is relevant as Smith Chart plots quite often become enclosures to design documents.

2.2.3 SWR and Γ

Specifications that define the desired performance characteristics for radio frequency devices require that such devices operate over a particular frequency band with an SWR less than some specified value. An SWR circle, constructed upon the Smith Chart, is used to show whether or not the device is performing to specification requirements. Impedance points lying anywhere within this SWR circle are said to be within specification limits whereas those outside the circle are not.

Figure 2-2 shows an SWR = 3:1 circle. The SWR of any particular impedance point may be determined by utilizing a compass to construct an arc of a circle, whose center is at the prime center, through the impedance point to intersect the resistance/conductance maximum axis. The point of intersection shows the desired SWR value. This process is illustrated in Figure 2-2 where the SWR of the 80 MHz impedance point is found to be 4.7:1.

The voltage reflection coefficient (Γ) is a complex quantity as noted in Equation (1-15) — having both magnitude and phase. The absolute magnitude, $|\Gamma|$, varies from zero at the chart center to unity at the chart rim. $|\Gamma|$ of any plotted impedance point may be derived by calculation using Equation (1-15) or may be determined by graphical construction from the chart as:

$$|\Gamma| = \frac{\text{distance from chart center to impedance point}}{\text{distance from chart center to chart rim}} \qquad (2\text{-}3)$$

2.2.4 Peripheral Scales

The peripheral scales that are shown on the rim of the Smith Chart are the Wavelength Scales and the Angle of Reflection Coefficient Scale.

The Wavelength Scales are labled "Towards Generator" and "Towards Load." These identify the direction of travel when it is desired to transform impedance values from one position on a transmission line to another. A travel of one complete circle on the chart represents a distance of one-half wavelength ($\lambda/2$) along the transmission line. Impedance/admittance relationships repeat each half wavelength along a uniform lossless (negligible attenuation) line. The zero point on the Wavelengths Scale may, therefore, be considered as any multiple of a half wavelength as well as zero. In performing impedance/admittance transformations, the electrical length of travel must be specified in terms of wavelengths. Equations (1-1) and (1-2) describe calculation of wavelength in both free space and in a dielectric filled line.

The Angle of Reflection Coefficient Scale provides the phase angle (ϕ) of the voltage reflection coefficient. ϕ is found by constructing a straight line from the chart center through the impedance point to intersect the Angle of Reflection Coefficient Scale on the chart rim. The Angle of Reflection Coefficient Scale can only be used with the impedance coordinates.

2.3 CARTER CHART

2.3.1 Coordinate Grid System

The Carter Chart, more commonly known as the Z-0 chart, permits the plotting of impedances in the polar form. The format that is shown in Figure 2-3 is a 50 ohm chart and is probably the more popular form in present day useage. The chart is not normalized and is used for plotting impedances where the output data provided from the measurement equipment utilized is in the form of magnitude (ohms) and phase (degrees).

The values shown on the vertical axis are in ohms. Positive phase angles are found on the right side of the chart whereas negative phase angles are on the left side.

There are some Carter Charts that are normalized. The chart shown in Figure 2-3 may be normalized by simply dividing all values on the vertical axis by 50 ohms such that the prime center has a value of 1.0. Normalized admittances, as in the case of the Smith Chart, would then appear diametrically opposite their respective impedance points.

Determination of SWR and Γ of any plotted impedance point may be found by graphical construction in a like manner as described for the Smith Chart.

2.3.2 Peripheral Scale

There is only one peripheral scale on the Carter Chart shown in Figure 2-3 and that is the Wavelengths Toward Load Scale. The process of impedance matching starts with a knowledge of the terminal impedance properties of the device to be compensated. Generally the impedance reference plane is at the bridge terminals of the measurement equipment with which the Carter Chart is used. Impedance transformation is, therefore, in only one direction and that is towards the load — hence the lack of need for a Towards Generator Wavelength Scale.

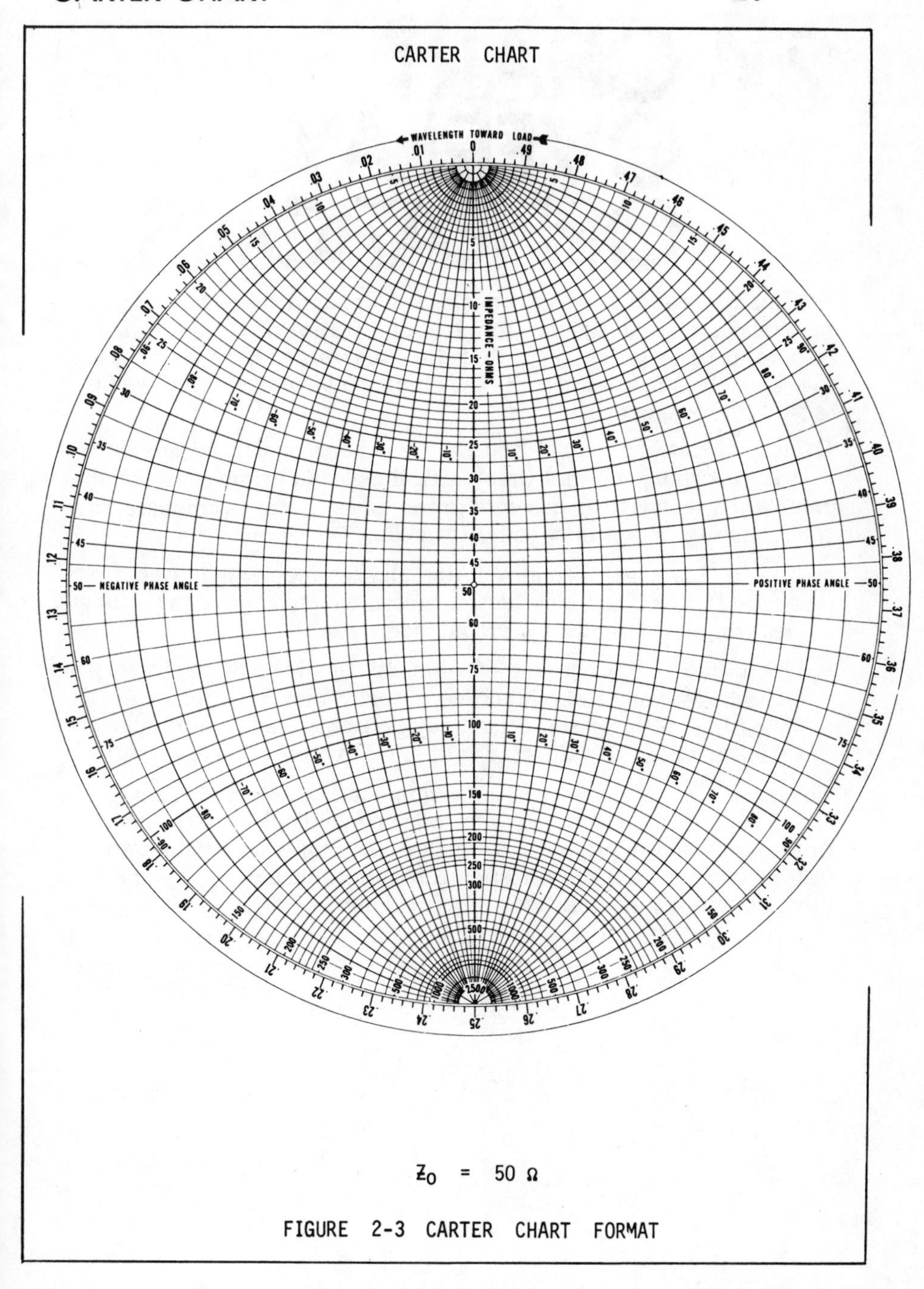

$Z_0 = 50\ \Omega$

FIGURE 2-3 CARTER CHART FORMAT

2.4 CHART OVERLAY OPERATIONS

Smith and Carter Charts are considered standard stockroom items at facilities engaged in the development and testing of radio frequency devices. It is the general case that the diameters of both charts are the same. This feature permits the charts to be used as overlays in several interesting operations.

Overlay operations require that the two charts be aligned one on top of the other such that their vertical axis and prime centers coincide. When the vertical axis zero points of both charts coincide, the following operations may be performed:

- If both charts are normalized, impedances appearing in polar form on the Carter Chart may be transposed to impedances in rectangular form on the Smith Chart and vice versa.
- When the Carter Chart is a 50 ohm chart and the Smith Chart prime center is assigned to be 50 ohms, actual values of impedances appearing on the Carter Chart may be transposed to normalized values on the Smith Chart. Figure 2-4 illustrates this process.

The other case to consider is the alignment of the vertical axis of the charts such that the zero point on one chart coincides with the infinity point on the other chart. In this case, impedance values appearing on one chart are transposed to admittance values on the other chart for the conditions described above.

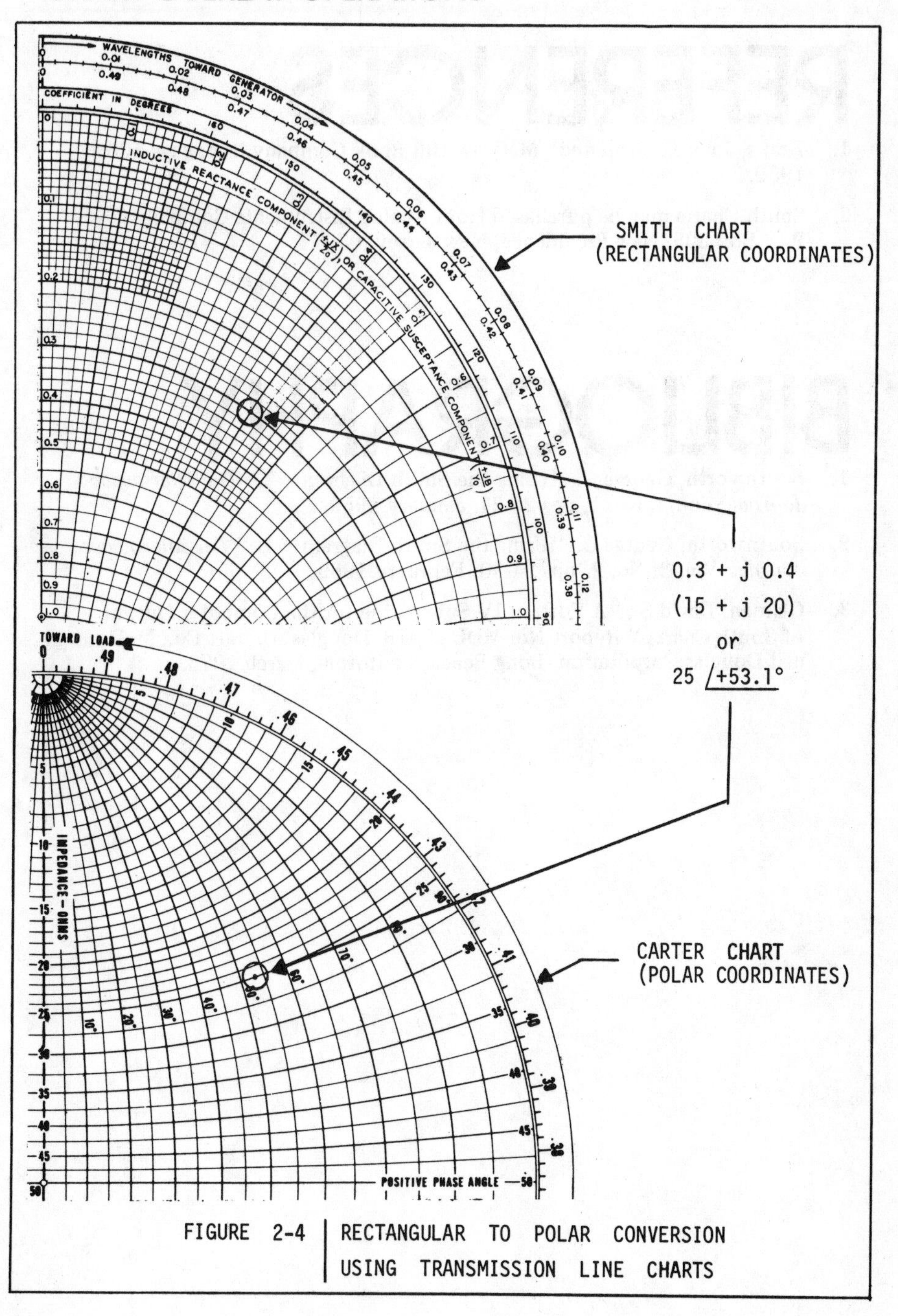

FIGURE 2-4 | RECTANGULAR TO POLAR CONVERSION USING TRANSMISSION LINE CHARTS

REFERENCES

1. Kraus, John D., *Antennas*, McGraw-Hill Book Company, Inc. Page 506, 1950.

2. Smith Charts may be purchased from Analog Instruments Company, P.O. Box 808, New Providence, New Jersey, 07974.

BIBLIOGRAPHY

1. Southworth, George C., "Using the Smith Diagram — I," *The Microwave Journal*, Vol. 2, No. 1, pps 25-31, January 1959.

2. Southworth, George C., "Using the Smith Diagram — II," *The Microwave Journal*, Vol. 2, No. 2, pps 24-30, February 1959.

3. Gelman, David S., "A Fortran IV Subroutine for the Computer Output of Smith Charts," Report No. MDC J5918, Douglas Aircraft Co., McDonnell Douglas Corporation, Long Beach, California, March 1973.

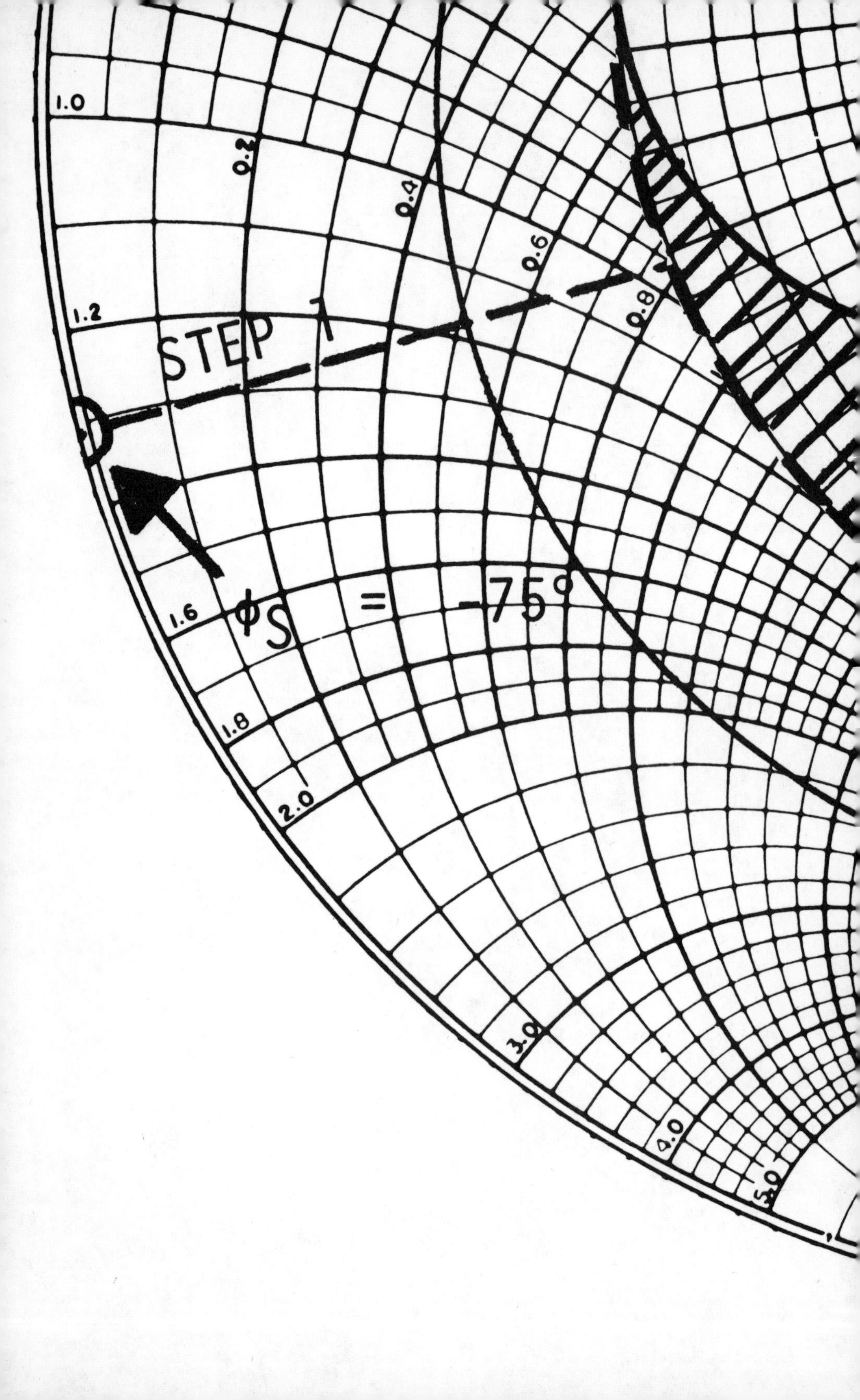
1.0
0.2
0.4
0.6
0.8
1.2
STEP 1
1.6
ϕ_S = -75°
1.8
2.0
3.0
4.0
5.0

3 TRANSMISSION LINE CHARTS—2

3.1 INTRODUCTION

Section II has presented a brief description of the coordinate grid system and peripheral scales of the Smith and Carter transmission line charts. The normalization process was discussed and a typical impedance/admittance plot given to illustrate the manner in which these quantities are entered upon the chart and the method of graphical determination of SWR for a given impedance point. In addition, it was shown that these two charts may be used in various overlay configurations to transpose impedance values in polar form to rectangular form and impedances on one chart to admittances on the other chart.

This section continues with the discussion of transmission line charts to describe fundamental operations that may be performed. Such operations include (a) impedance transformation on "lossless", "lossy", and non-uniform lines, (b) input impedance properties of open and shorted line segments, and (c) chart operations in the solution of a simple network.

3.2 IMPEDANCE TRANSFORMATION

Impedance transformation, as discussed herein, is the chart operation of transforming a given set of impedance data from one position on a transmission line to another. Such operations are frequently used in matching network analysis where it is desired to position the impedance curve to obtain the maximum effectiveness of a particular compensating network.

3.2.1 Line Travel ($\theta \ell$)

Transformation of impedances or admittances along a transmission line involves primarily the use of the peripheral Wavelengths Scales in the direction either towards the source or towards the load. Line travel must be specified in terms of wavelengths. A given physical length of transmission line has one electrical length at one frequency and a different electrical length at another frequency. If the electrical length ($\theta \ell$) is known at one frequency, then the electrical lengths at other frequencies may be readily determined by ratio for TEM lines from:

$$\theta \ell_2 = \frac{f_2}{f_1} \cdot \theta \ell_1 \qquad (3\text{-}1)$$

Where: $\theta \ell_2$ = electrical length at frequency f_2 (wavelengths)
$\theta \ell_1$ = electrical length at frequency f_1 (wavelengths)

As an example consider a rexolite ($\epsilon = 2.5$) filled coaxial line whose physical length is 100 inches and it is desired to determine the electrical length in the 30 MHz to 75 MHz frequency band. The wavelength in the dielectric medium (λm), at 30 MHz may be found from Equations (1-1) and (1-2) as:

$$\lambda m = \frac{\lambda o}{\sqrt{\epsilon}} = \frac{(300)\,(39.4)}{(30)\,(\sqrt{2.5})} = 249.0 \text{ inches}$$

and the electrical length ($\theta\ell$) of the coaxial line at this frequency is:

$$\theta\ell_{30} = \frac{100''}{249''} = 0.402\lambda$$

The electrical lengths, at other frequencies within the band of interest, may be found by applying Equation (3-1) and are as follows:

frequency (MHz)	$\theta\ell$ *(wavelengths — λ)*
30.0	0.402
40.0	0.536
50.0	0.670
60.0	0.804
70.0	0.936
75.0	1.001

Since impedance relationships repeat each half-wavelength (λ/2) along a uniform "lossless" line, the line travel in the above example would be:

frequency (MHz)	$\theta\ell$ *(wavelengths — λ)*
30.0	0.402
40.0	0.036
50.0	0.170
60.0	0.304
70.0	0.436
75.0	0.001

3.2.2 "Lossless" Line Case

A "lossless" transmission line is one that is characterized by negligible attenuation. Equation (1-8) defines the characteristic impedance of such a line in terms of inductance and capacitance per unit length. Figure (3-1) shows a load whose impedance at 30 MHz and 40 MHz is $Z_{30} = 15 + j\,10\Omega$ and $Z_{40} = 10 + j\,25\Omega$. Let it be desired to transform this impedance data to Reference Plane B-B which represents a line travel of .085λ at 30 MHz. The process is described in a step-by-step manner as:

1/Normalize Z_{30} and Z_{40} to $Z_o = 50\Omega$ as $Z'_{30} = 0.3 + j\,0.2$ and $Z'_{40} = 0.2 + j\,0.5$. These values are shown plotted in the figure and identified as the impedance curve at Reference Plane A-A (Z'_{AA}).

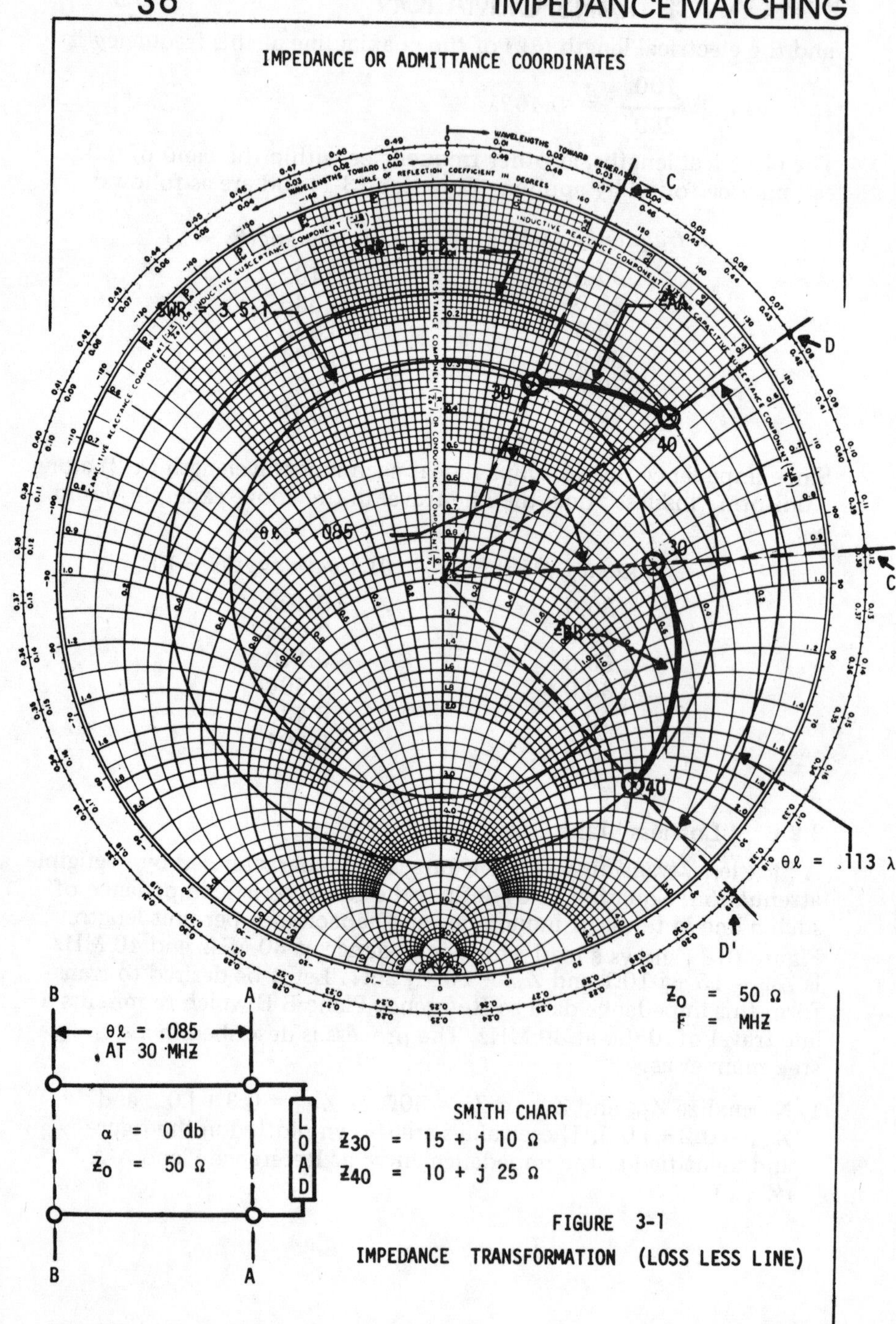

FIGURE 3-1
IMPEDANCE TRANSFORMATION (LOSS LESS LINE)

2 /For a "lossless" line, impedances transform on constant SWR circles. Construct these SWR circles to include Z'_{30} and Z'_{40} on the Z'_{AA} impedance curve.

3 /Construct radial lines from the chart center through the impedance points to intersect the Wavelengths Scales. C (.034λ) and D (.076λ) are the points of intersection and define the initial starting positions for line travel.

4 /Line travel at 30 MHz is .085λ and at 40 MHz [from Equation (3-1)] is 0.113λ. End positions, C′ and D′, of line travel are found by adding the line travel at each frequency to the starting position values as:

$$C' = .034\lambda + .085\lambda = 0.119\lambda$$
$$D' = .076\lambda + .113\lambda = 0.189\lambda$$

5 /Construct radial lines from the chart center to C′ and D′. The points of intersection of these lines and the SWR circles define the impedance values at Reference Plane B-B. These values are:

$$Z'_{30} = 0.5 + j\,0.8 \quad , Z_{30} = 25 + j\,40\Omega$$
$$Z'_{40} = 1.0 + j\,2.1 \quad , Z_{40} = 50 + j\,105\Omega$$

Thus the impedance curve Z'_{BB} represents the impedance characteristics of the load as transformed to Reference Plane B-B.

3.2.3 "Lossy" Line Case

Figure (3-2) illustrates transformation along a line containing attenuation. Load conditions and line travel are the same as that given in Figure (3-1). Transformation on a "lossy" line is accomplished in the same manner as described in 3.2.2 for the "lossless" line case except that the values obtained at the end of line travel must be corrected for attenuation. The transformed impedance points marked thus (★) in Figure (3-2) are for the "lossless" line case as previously described. Correction for attenuation at the end of line travel involves the following steps:

1/Calculate the reflection coefficients of the 30 MHz and 40 MHz impedances at Reference Plane A-A utilizing Equation (1-16) as:

$$|\Gamma_{AA}|_{30} = \frac{SWR - 1}{SWR + 1} = \frac{3.5 - 1}{3.5 + 1} = 0.566$$

$$|\Gamma_{AA}|_{40} = \frac{SWR - 1}{SWR + 1} = \frac{6.2 - 1}{6.2 + 1} = 0.723$$

2/Convert the line attenuation that is given in db (α = 2db) to nepers utilizing Equation (1-21) as:

$$\alpha = 2 \text{ db} = \frac{2}{8.68} = 0.231 \text{ nepers}$$

$$2\alpha = 0.462 \text{ nepers}$$

3/Determine the reflections coefficients at Reference Plane B-B from Equation (1-19) as:

$$|\Gamma_{BB}|_{30} = |\Gamma_{AA}|_{30} \cdot \epsilon^{-2\alpha}$$

$$= (0.566) \cdot \epsilon^{-0.462} = 0.357$$

$$|\Gamma_{BB}|_{40} = |\Gamma_{AA}|_{40} \cdot \epsilon^{-2\alpha}$$

$$= (0.723) \cdot \epsilon^{-0.462} = 0.455$$

4/The input SWR of the 30 MHz and 40 MHz impedance points at Reference Plane B-B, using Equation (1-16), is:

$$SWR_{30} = \frac{1 + |\Gamma_{BB}|_{30}}{1 - |\Gamma_{BB}|_{30}}$$

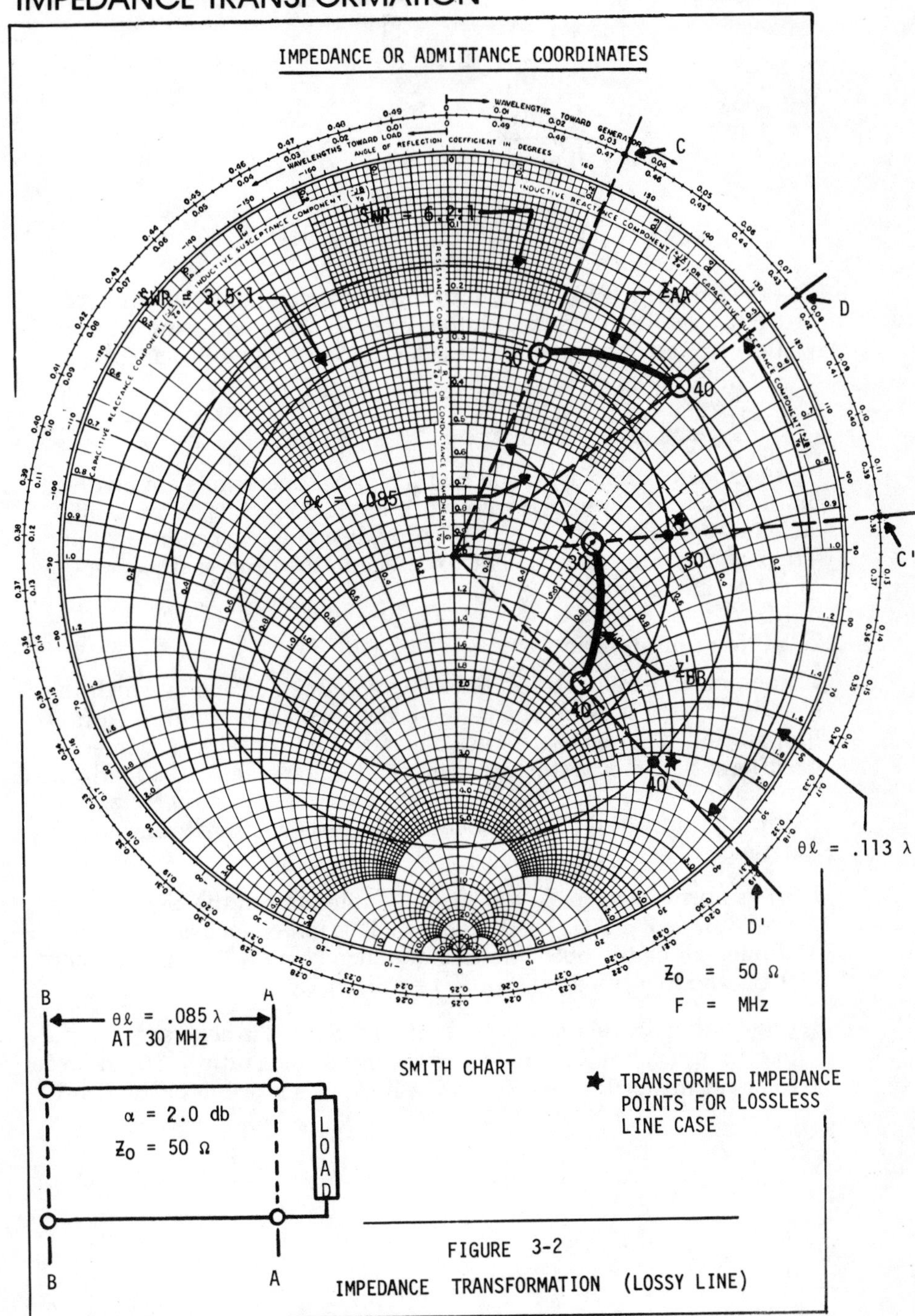

FIGURE 3-2
IMPEDANCE TRANSFORMATION (LOSSY LINE)

$$= \frac{1 + 0.357}{1 - 0.357} = 2.12:1$$

$$SWR_{40} = \frac{1 + |\Gamma_{BB}|_{40}}{1 - |\Gamma_{BB}|_{40}}$$

$$= \frac{1 + 0.455}{1 - 0.455} = 2.67:1$$

5/Construct the SWR_{30} circle to intersect radial line OC′ and the SWR_{40} circle to intersect radial line OD′. The points of intersection are the impedance values at Reference Plane B-B and are:

$Z'_{30} = 0.72 + j\,0.6, \quad Z_{30} = 36 + j\,30\ \Omega$

$Z'_{40} = 1.4 + j\,1.15, \quad Z_{40} = 70 + j\,57.5\ \Omega$

In the above example, several mathematical operations are involved in the determination of the network input SWR. Charts are available in literature[1], however, that provide a graphical solution and are probably more convenient for usage in the process of correction for line attenuation.

3.2.4 Non-uniform Line

A non-uniform transmission line is here defined as one in which the electrical characteristics change with distance along the direction of propagation. Figure (3-3) shows one such line containing segments of different characteristic impedance. At Reference A-A, a load whose impedance $Z_L = 20 + j\,30\Omega$ is given at a frequency of 30 MHz. The following is involved in the process of impedance transformation to Reference Plane D-D:

1/Z_L is normalized to $Z_0 = 50\Omega$ and entered upon the chart of Figure (3-3) as $Z'_{AA\,(50)} = 0.4 + j\,0.6$. This impedance is next transformed along the 50Ω line to Reference Plane B-B, a line travel of 0.1λ, and identified as $Z'_{BB(50)} = 1.5 + j\,1.6$.

2/Prior to line travel on the $Zo = 35\Omega$ section, it is necessary to multiply $Z'_{BB\,(50)}$ by the characteristic impedance ratio $^{50}/_{35}$ in order to be normalized to the $Zo = 35\Omega$ line. This is accomplished and

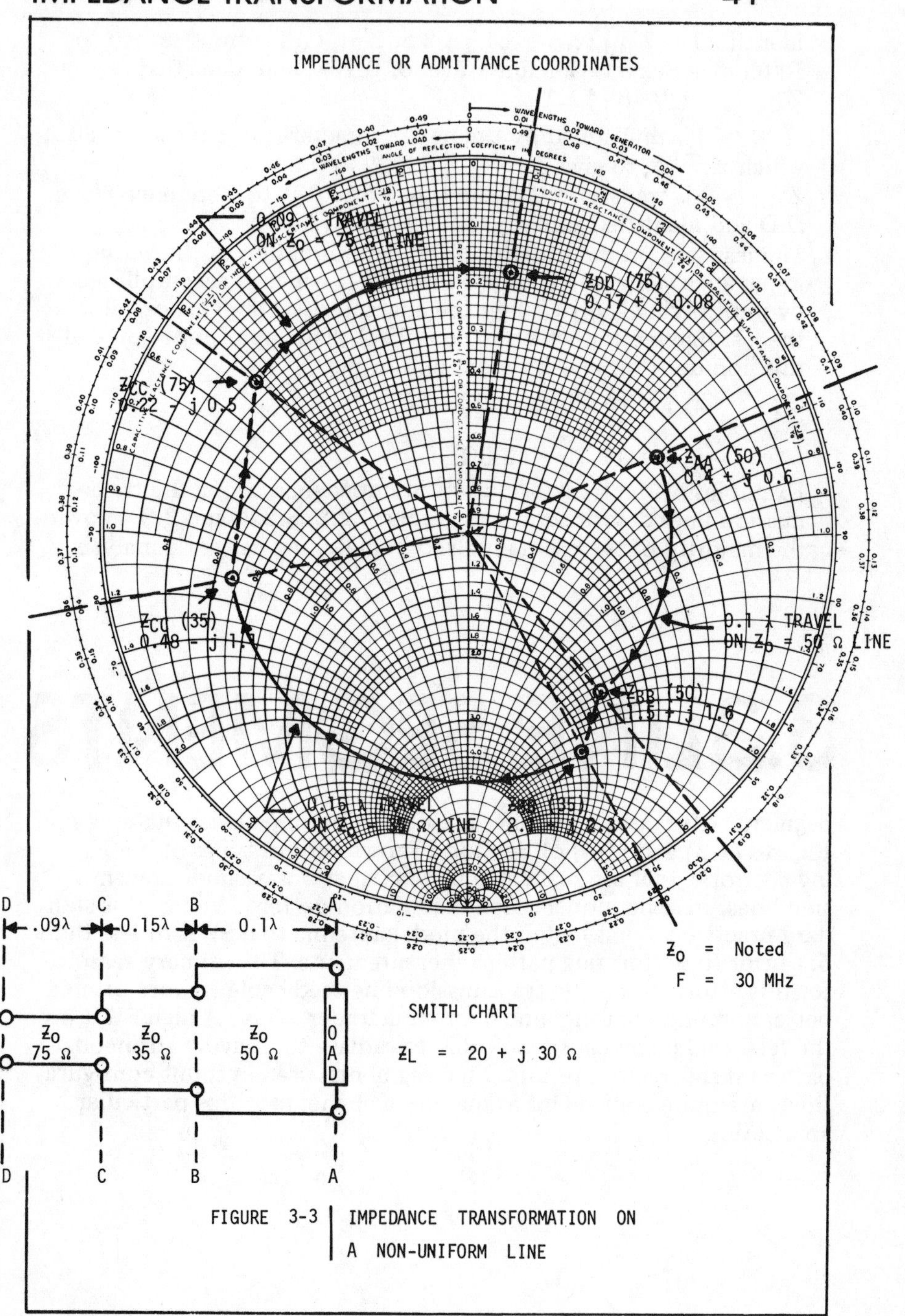

FIGURE 3-3 | IMPEDANCE TRANSFORMATION ON A NON-UNIFORM LINE

identified as $Z'_{BB}(_{35}) = 2.1 + j\,2.3$. $Z'_{BB}(_{35})$ is transformed to Reference Plane C-C, a line travel of 0.15λ, and identified as $Z'_{CC}(_{35}) = 0.48 - j\,1.1$.

3/ $Z'_{CC}(_{35})$ is multiplied by the next characteristic impedance ratio, which is $^{35}/_{75}$, to give $Z'_{CC}(_{75}) = 0.22 - j\,0.5$.
$Z'_{CC}(_{75})$ is transformed a line travel of 0.09λ to Reference Plane D-D and identified as $Z'_{DD}(_{75}) = 0.17 + j\,0.08$.

4/ The last step, in the process, is the determination of the actual impedance value at Reference Plane D-D which is accomplished by multiplying $Z'_{DD}(^{75})$ by the characteristic impedance value of the last line section which is $Z_o = 75\ \Omega$. The input impedance of this non-uniform line is thus found to be:

$$Z_{DD} = (0.17 + j\,0.08) \cdot 75$$

$$= 12.75 + j\,6.0\Omega$$

Examples of the transformation of impedances along the transmission lines described in paragraphs 3.2.2, 3.2.3, and 3.2.4 cover the majority of the cases involved in matching network analysis.

3.3 LINE SEGMENTS

Segments of transmission lines are utilized in impedance matching. Figures (1-3) and (1-4) of Section I shows various types of matching networks that are realizable in coaxial and waveguide transmission lines. In scale model antenna radiation pattern studies, it is standard practice to first match the model antenna to a SWR of less than 5:1 prior to performing pattern measurements. The primary reason for this effort is that the transmission line feed cable is more often a better antenna than the model antenna under study. A high SWR on the feed cable may cause the cable to radiate thus giving erroneous pattern measurement results. Line segments, used in shunt configurations, are quite convenient as matching elements in this particular application.

Equations (1-5) and (1-6) provide the input impedance of open circuited and short circuited line segments. The Smith Chart provides a ready graphical solution to the above equations. An example is shown in Figure (3-4) where a segment of open and shorted line is given. Since the input impedance of such segments is always reactive ($\pm j\chi$), the impedance curves are shown on the rim of the chart. The input impedance, at any position on the line segment, may be determined by multiplying the normalized value on the chart rim by the characteristic impedance assigned to the chart center. For example, let it be assumed that the chart center Zo = 75Ω and it is desired to examine the impedance at various positions along the 0.176λ shorted segment. Referring to Figure (3-4) the following is compiled:

Reference Plane	*Normalized Z*	*Actual Z*
A-A	$Z'_{AA} = +j\,0$	$Z_{AA} = +j\ 0\ \Omega$
B-B	$Z'_{BB} = +j\,0.4$	$Z_{BB} = +j\ 30\ \Omega$
C-C	$Z'_{CC} = +j\,0.9$	$Z_{CC} = +j\ 67.5\ \Omega$
D-D	$Z'_{DD} = +j\,2.0$	$Z_{DD} = +j\ 150\ \Omega$

Examination of the figure also shows that the input impedance of an λ/4 shorted cable is infinity ohms and of an open circuited cable is zero ohms. Also a shorted λ/8 cable has a normalized +j 1.0 input impedance whereas an λ/8 open circuited cable has a normalized -j 1.0 input impedance.

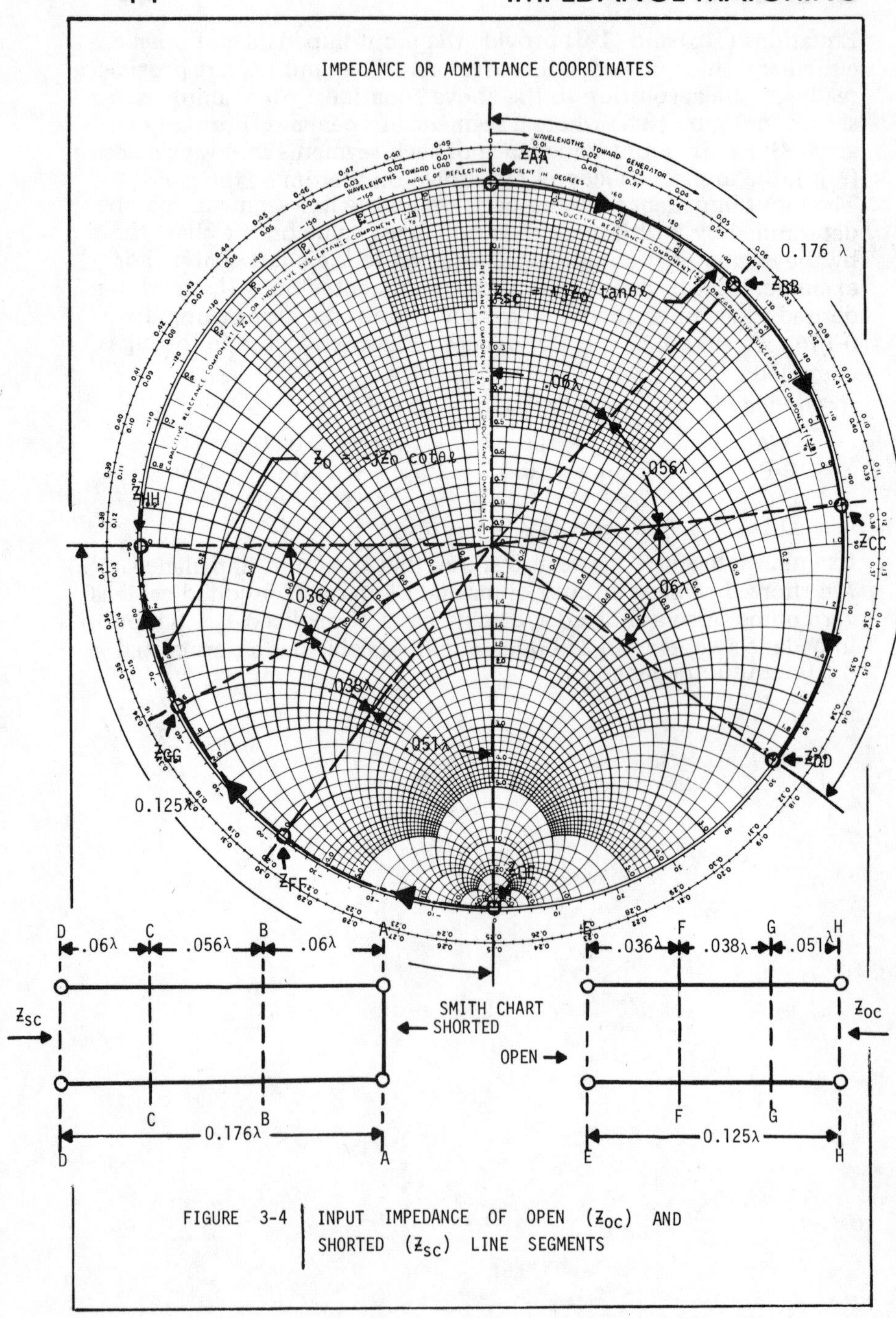

FIGURE 3-4 INPUT IMPEDANCE OF OPEN (Z_{OC}) AND SHORTED (Z_{SC}) LINE SEGMENTS

3.4 SERIES/SHUNT NETWORK

Graphical solutions to simple network problems are accomplished quite easily using the Smith Chart. In Figure (3-5), a lumped constant T-network is shown that is selected to match a given load whose impedance at 30 MHz is $Z_L = 5 + j\,0\Omega$. The load is pure resistance and a perfect match is sought. At higher frequencies, a quarter wave transmission line transformer section such as that shown in Figure (1-3a) could be considered for the matching network since the physical length of the section would probably not be objectionable. At 30 MHz, however, a $\lambda/4$ section in a coaxial rexolite line represents a physical length of 62 1/4 inches. The network shown in the figure is an equivalent quarter wave transformer utilizing lumped constant matching elements.

This example is selected to describe the method of handling series and shunt impedances in the process of network solution. A good rule to follow is to work with impedances when combining series elements and admittances when combining shunt elements. The solution to the network shown in Figure (3-5) to derive the input impedance Z_{DD} follows:

$1/Z_{Load}$ (Z_{AA}), Z_{L1}, Z_{L2}, and Z_{C1} are normalized and plotted as:

$$Z'_{AA} = 0.1 + j\,0$$
$$Z'_{L1} = Z'_{L2} = 0 + j\,0.32$$
$$Z'_{C1} = 0 - j\,0.32$$

Since C1 is a shunt element, the admittance Y'_{C1} (diametrically opposite Z'_{C1}) = 0 + j 3.1 is also plotted.

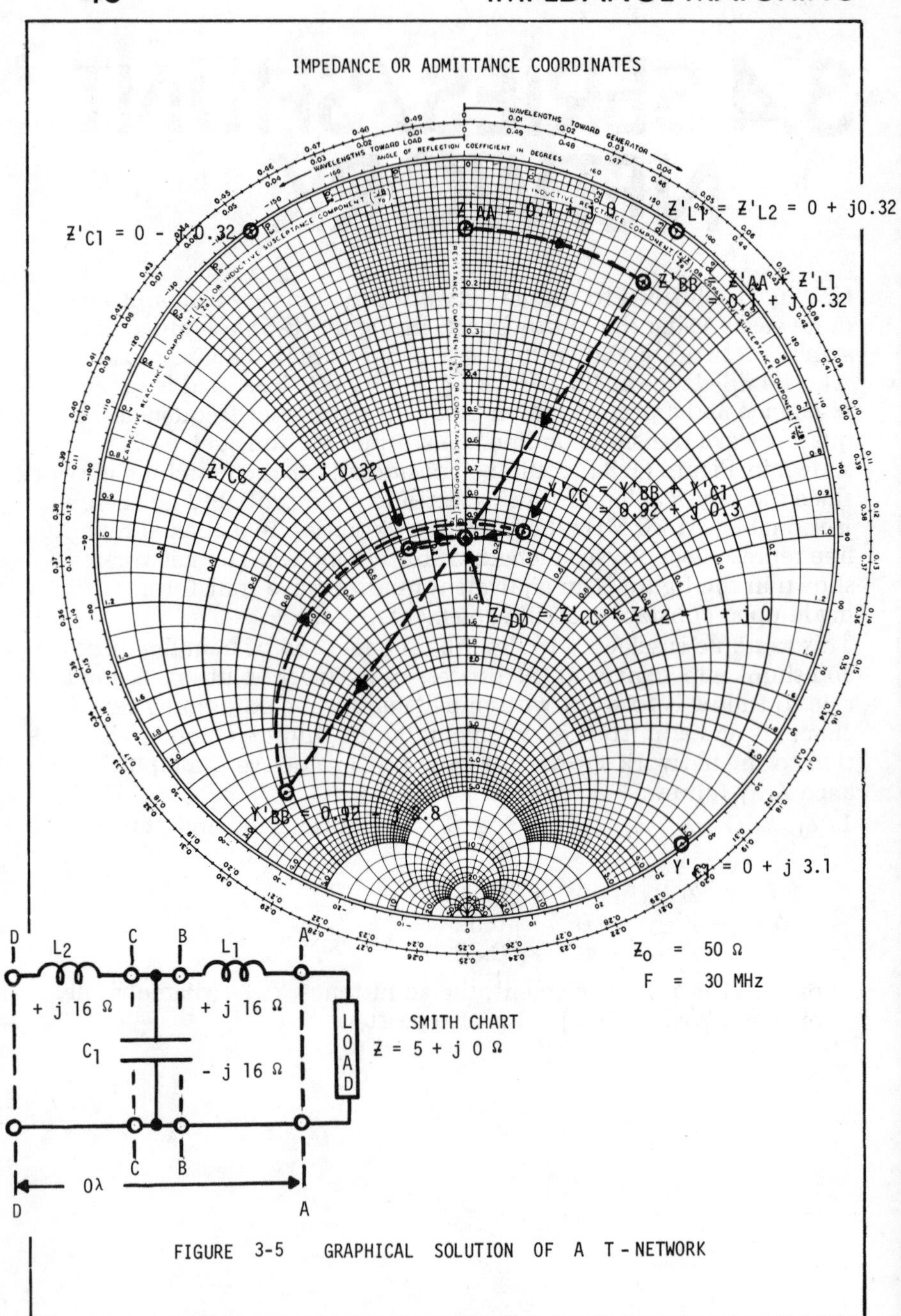

FIGURE 3-5 GRAPHICAL SOLUTION OF A T-NETWORK

2/The impedance at Reference Plane B-B is composed of series elements Z'_{AA} and Z_{L1} which is:

$$
\begin{aligned}
Z'_{BB} &= Z'_{AA} + Z'_{L1} \\
&= (0.1 + j\,0) + (0 + j\,0.32) \\
&= 0.1 + j\,0.32
\end{aligned}
$$

3/Y'_{BB} is diametrically opposite Z'_{BB} and is plotted as $Y'_{BB} = 0.92 - j\,2.8$. Y'_{BB} is added to Y'_{C1} to give $Y'_{CC} = 0.92 + j\,0.3$. $Z'_{CC} = 1 - j\,0.32$ is found diametrically opposite Y'_{CC}.

4/The desired input impedance at Reference Plane D-D is next found by adding Z'_{CC} and Z'_{L2} in series to give:

$$
\begin{aligned}
Z'_{DD} &= Z'_{CC} + Z'_{L2} \\
&= (1 + j\,0.32) + (0 + j\,0.32) \\
&= 1 + j\,0 = \text{perfect match} \\
Z_{DD} &= (1 + j\,0) \cdot 50 = 50 + j\,0\Omega
\end{aligned}
$$

3.5 ADDITIONAL REMARKS

Sections II and III have provided the reader with a fundamental knowledge of the two most popular forms of transmission line charts. These charts have become quite important to all practitioners in the field of radio frequency work and particularly to those engaged in impedance matching. An insight as to the degree of importance of the Smith Chart, for example, may be found in the following remark of Dr. George C. Southworth:[2]

> "In this writer's opinion, the Smith diagram has been one of the more important device contributions made to microwave technique during the last couple of decades."

REFERENCES

1. International Telephone and Telegraph Corporation, *Reference Data for Radio Engineers*, Fourth Edition, pages 570 and 571, 1957.

2. Southworth, George C., "More About Phil Smith and His Diagram," *The Microwave Journal*, Vol. 1, No. 2, pgs, 26-28, September 1958.

BIBLIOGRAPHY

1. Mathis, H.F., "Logarithmic Transmission Line Charts," *Electronics*, September 1960.

2. Blanchard, W.C., "A Do-It-Yourself Transmission Line Impedance Chart," *Microwaves*, pages 46-52, April 1969.

3. Markin, Joseph, "Smith Chart Applications," Tele-Tech & Electronic Industries, pp 85-88, May 1953.

4. Cholewski, L.S., "Some Amateur Applications of the Smith Chart," QST., pp 28-31, January 1960.

4 IMPEDANCE MEASUREMENT METHODS

4.1 INTRODUCTION

Impedance matching analysis must first start with a knowledge of the terminal impedance properties of the r.f. device to be compensated. It has been proven by theory and established in practice that greater impedance bandwidths may be achieved when the matching network is incorporated as near as practical to the device terminals. An inherent characteristic of any impedance curve, as may be shown on the Smith Chart, is that the impedance points tend to "spread" as the electrical length from the device terminals is increased. The spreading of the impedance points has the meaning that the bandwidth potential has been reduced. It is, therefore, of importance that one entire section of this book be devoted to a summary discussion of various measurement methods in present day usage that describe the terminal impedance properties of r.f. devices.

The more frequently used impedance measurement methods are (a) the VHF Bridge, (b) the Standing Wave Detector, and (c) the Slotted Line. The HP 803A VHF Bridge, for example, may be used at frequencies from 50 MHz to 500 MHz, whereas the Slotted Line finds applications from 500 MHz throughout the microwave frequency spectrum. The PRD 219 Standing Wave Detector is an instrument that permits impedance measurements within the 20 MHz to 2000 MHz frequency range. These methods are employed to measure devices that radiate r.f. energy such as antennas and also devices that do not radiate such as filters, hybrids, terminations, attenuators, and so forth. There are other methods of measuring impedance using such instrumentation as "Q" Meters, Vector Impedance Meters, and swept frequency equipment; however, the reader is referred to the selected bibliography given at the end of this section for a description of these methods.

This section is more primarily concerned with the type of data obtained from the above noted measurement methods and the technique of data reduction rather than in operational procedures. Specific equipment operational procedures may be found in literature.[1,2,3]

4.2 IMPEDANCE REFERENCE PLANE

In Section II, it was noted that impedance/admittance data plotted upon the Smith Chart is meaningless if such data is not properly identified. Part of the proper identification requires that the Impedance Reference Plane be defined. The Impedance Reference Plane simply identifies that position, either on the transmission line or at the device terminals, for which the data plotted upon the Smith Chart is valid.

It is general practice in performing impedance measurements to establish the Impedance Reference Plane, as close as practical to the device terminals. Broader impedance bandwidths may be achieved when matching networks are incorporated at or near this location. Figure 4-1 shows the preferred location of the Impedance Reference Plane for differing measurement test setups. Establishment of this

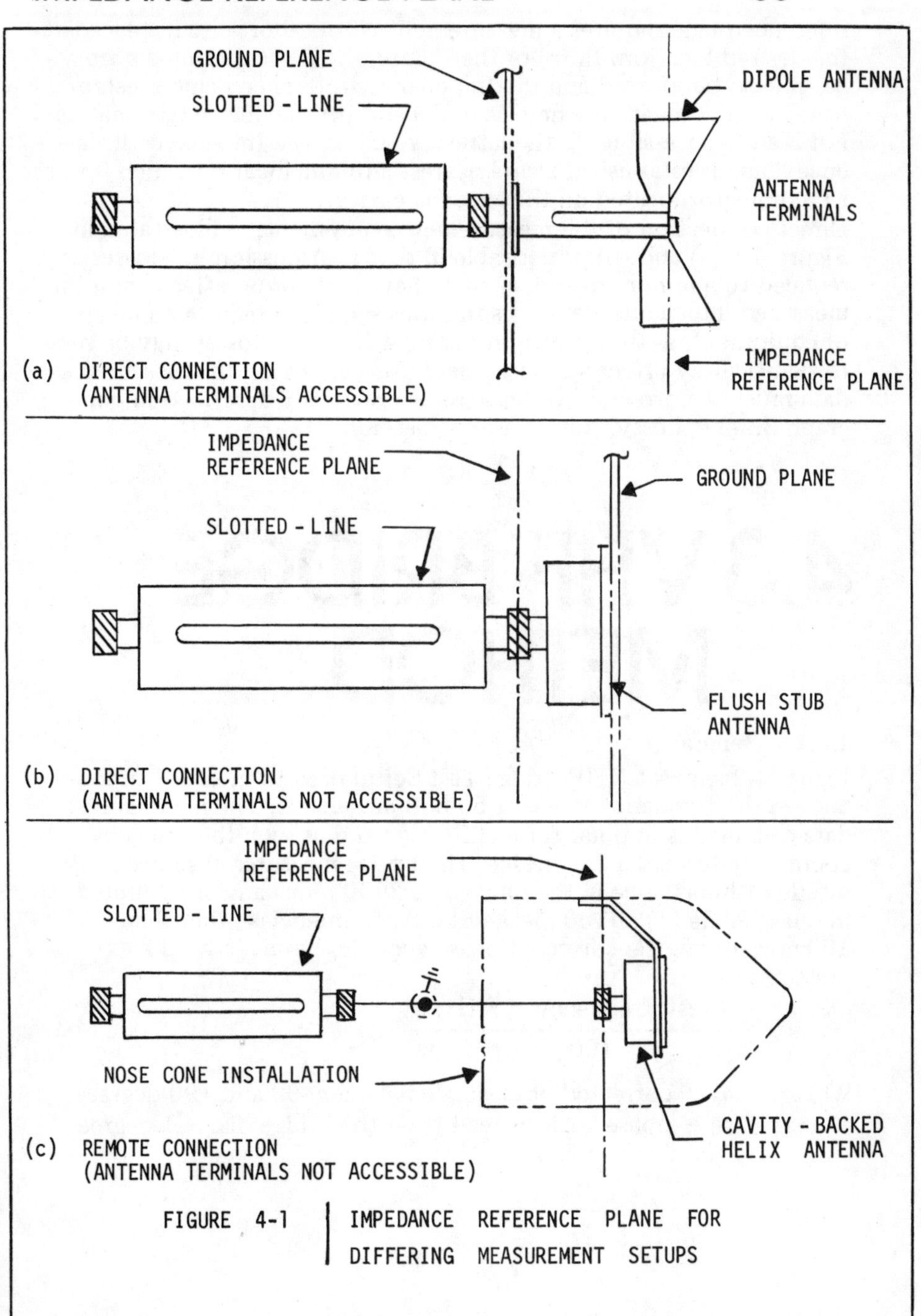

FIGURE 4-1 | IMPEDANCE REFERENCE PLANE FOR DIFFERING MEASUREMENT SETUPS

reference plane requires a measurement with a short circuit placed at the desired location. In (a) of the Figure, the device terminals are accessible to be shorted and the Impedance Reference Plane is established at that location where as in (b) and (c) the device terminals are not readily accessible. In the latter two cases, the Impedance Reference Plane is established at the nearest antenna location which is at the r.f. connector located on the antenna cavity.

Direct connection of the measurement equipment, as illustrated in Figure 4-1 (a) and (b), is desirable in that transmission line losses are reduced to a minimum and, as such, have little or no affect upon the measured impedance data. In situations requiring remote connection of equipment, as that shown in Figure 4-1(c), line losses may be such to significantly affect measured data. In these cases, the measured data must be corrected for line loss as previously described in paragraph 3.2.3 ("Lossy" Line Case) of Section III.

4.3 VHF BRIDGE METHOD

4.3.1 General

Figure 4-2 shows a VHF Bridge Test Setup used to measure impedance in the frequency range of 50-500 megahertz. The impedance data obtained is in polar form ($|Z| \angle \pm \theta$) and as such the bridge is commonly termed a Z-θ Bridge. The bridge measures absolute magnitude of impedance in the range of 2-2000 ohms and is calibrated in phase angle (θ) at 100 megahertz. At frequencies other than 100 megahertz, the corrected phase angle (θ_c) is given by the expression:

$$\theta_c = \frac{(\text{Test Frequency} - \text{MHz})}{100} (\theta_d) \qquad (4\text{-}1)$$

Where: θ_c = corrected phase angle between -90 and +90 degrees.
θ_d = phase angle as read from the bridge dial — ± degrees.

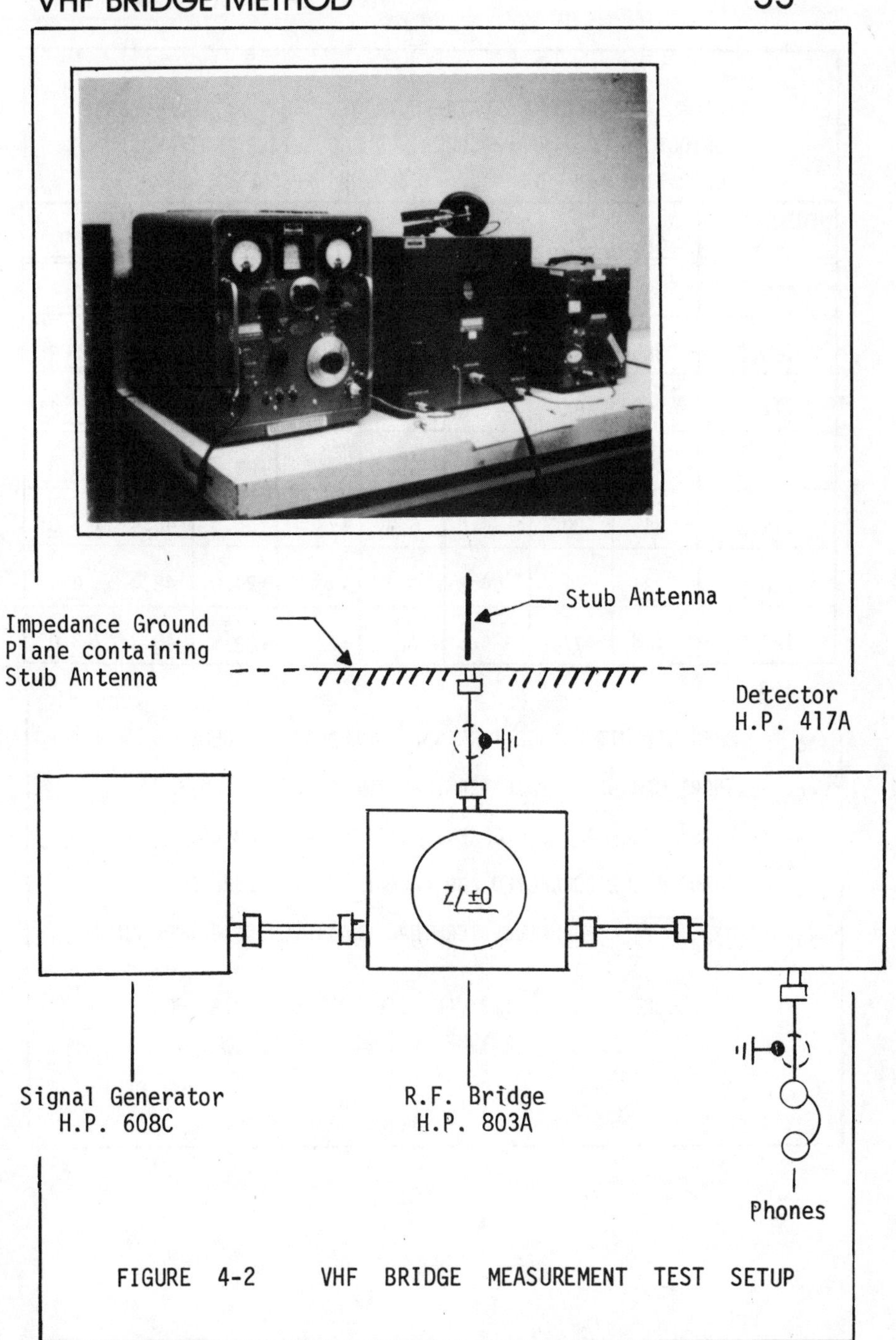

FIGURE 4-2 VHF BRIDGE MEASUREMENT TEST SETUP

TYPICAL CALIBRATION DATA OBTAINED FOR AN H.P. 803A BRIDGE USED TO MEASURE VOR / LOCALIZER ANTENNAS OPERATING IN THE 108 - 122 MHZ FREQUENCY BAND:

FREQUENCY MHZ	Z_{OC}	θ_d	θ_c	Z_{SC}	θ	θ_c	$\sqrt{Z_{OC} \cdot Z_{SC}}$	
108.0	381	-87	-93.8	6.2	+81	+87.5	48.6	$\angle -3.1°$
109.6	386	-85	-93.1	6.35	+80	+87.7	48.9	$\angle -2.7°$
112.0	365	-84	-94.0	6.5	+78	+87.5	48.7	$\angle -3.2°$
116.6	350	-81	-94.5	6.6	+75	+87.5	48.1	$\angle -3.0°$
118.0	345	-80	-94.5	6.8	+75	+88.5	48.4	$\angle -3.0°$
120.0	335	-79	-94.8	7.0	+79	+94.8	48.4	$\angle 0°$
122.0	328	-77.5	-94.5	7.1	+72.5	+88.5	48.3	$\angle -3.0°$

Z_{OC} = OPEN CIRCUIT BRIDGE TERMINAL IMPEDANCE - OHMS

Z_{SC} = SHORT CIRCUIT BRIDGE TERMINAL IMPEDANCE - OHMS

θ_d = PHASE ANGLE AS READ ON BRIDGE DIAL - DEGREES

θ_c = PHASE ANGLE CORRECTED FOR FREQUENCY - DEGREES

$Z_0 = \sqrt{Z_{OC} \cdot Z_{SC}}$ = BRIDGE TERMINAL CHARACTERISTIC IMPEDANCE

FIGURE 4-3 | TYPICAL CALIBRATION DATA FOR A VHF BRIDGE

The uncorrected accuracy of the measurement is 5% to 6% in magnitude and 3° to 4° in phase angle. These accuracies are usually acceptable for the majority of measurement tasks encountered in general impedance development work. Correction curves are provided with the equipment, however, that may be used to increase the measurement accuracy to 2% in magnitude and to 1.2° in phase for those cases requiring greater accuracy.

4.3.2 Equipment Calibration

The validity of test data is generally supported by measurements that establish that the test equipment was within proper calibration. Such measurements form a natural part of the test documentation. Figure 4-3 shows typical calibration data obtained for a VHF Bridge used to measure the impedance of VOR/Localizer antennas operating in the frequency band of 108-122 MHz. The data was obtained for open and short circuited conditions at the bridge terminals. The principal results are given in the last column of the figure. The characteristic impedance (Zo) of the bridge terminals was determined from the measured data using Equation (1-10). For example, the Zo of the bridge terminal at 108 MHz was found to be:

$$
\begin{aligned}
Zo_{(108)} &= \sqrt{Zoc \cdot Zss} \\
&= \sqrt{(381 \angle -93.8^\circ)(6.2 \angle +87.5)} \\
&= \sqrt{2360 \angle -6.3^\circ} \\
&= 48.6 \angle -3.1^\circ
\end{aligned}
$$

Examination of the results shows good agreement with the published bridge accuracies of 5% in magnitude and 3° in phase.

4.3.3 Data Reduction

Bridge measurement data for a slot antenna operating in the 210-240 MHz band is given in Figure 4-4. The particular antenna shown is a cavity backed single folded slot in which partial impedance compensation is achieved by the selection of appropriate characteristics for

(a) SLOT ANTENNA

(b) FEED TERMINAL DETAILS

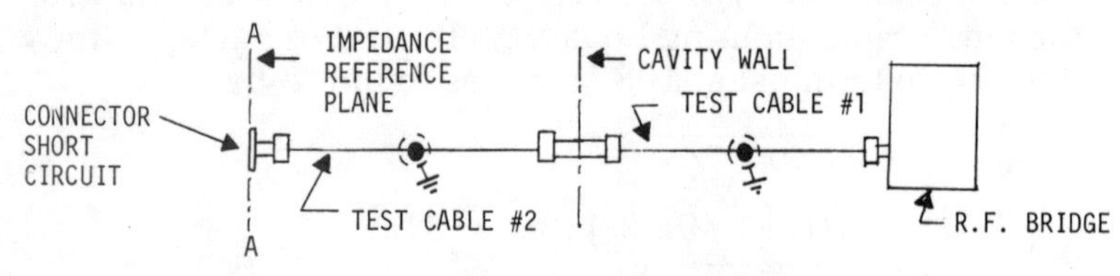

(c) TEST CABLE ARRANGEMENT

(d) CABLE SHORT MEASUREMENTS

FREQUENCY (MHZ)	$\|Z\|$ (Ω)	θ (DEG)	θ_c (DEG)	$\theta \ell$ (λ)
210.	200.	-40.5	-85.5	.289
220.	69.	-40.0	-88.0	.350
230.	30.	-38.0	-87.4	.414
240.	8.	-35.0	-84.0	.475

(e) SLOT ANTENNA MEASUREMENTS

FREQUENCY (MHZ)	$\|Z\|$ (Ω)	θ (DEG)	θ_c (DEG)	$\theta \ell$ (λ)
210.	60.	-37	-77.7	.289
220.	9.2	-22	-48.5	.350
230.	31.	+33	+75.9	.414
240.	105.	+31	+74.4	.475

FIGURE 4-4 BRIDGE MEASUREMENT DATA FOR A SLOT ANTENNA

the folded and driven slots. The set of measurements is selected to illustrate the method of data reduction and the plotting of results on the Carter Chart. The data reduction procedure follows:

1/Impedance Reference Plane (Z_{AA}) — Z_{AA} is established by obtaining a set of cable short measurements [Figure 4-4(d)] with a connector short circuit placed at the antenna terminals. The purpose of these measurements is to determine the electrical length ($\theta\ell$) of the transmission line from the bridge measurement plane to Z_{AA} in terms of wavelengths.

2/Finding $\theta\ell$ — Cable short measurements [$|Z| \angle \pm\theta c$] of Figure 4-4(d) are plotted upon the Carter Chart of Figure 4-5. Through each plotted point, a radial line is constructed to intersect the peripheral wavelengths scales. For example, the radial line through the 210 MHz point intersects the wavelengths scale at 0.211 λ. $\theta\ell$ is found by subtracting this value from 0.5 λ as:

$$\theta\ell_{(210)} = 0.5\,\lambda - 0.211\,\lambda = 0.289\,\lambda$$

The electrical lengths of the transmission line at the remaining frequencies are found in a like manner.

3/The slot antenna measurements [$|Z| \angle \pm\theta c$] of Figure 4-4(e) are next plotted in Figure 4-6 and corrected for the above values of $\theta\ell$ to provide the impedance curve Z_{AA}. For simplicity, construction is shown only for the 210 MHz impedance data in the figure. The graphical construction start with a radial line through the measured $Z_{(210)}$ impedance point to intersect the wavelengths scale at an initial starting position (θi) for line transformation. To θi is added $\theta\ell$ to give the transformed position (θt) on the wavelengths scale. In this case, θt is:

$$\theta t = \theta i + \theta\ell$$
$$= 0.14\,\lambda + 0.289\,\lambda = 0.429\,\lambda$$

A radial line is next constructed from the chart center to intersect θt. The antenna terminal impedance, at 210 MHz, lies on this radial line, at the same distance from the chart center as for the initial measured plotted impedance point. The construction assumes the "lossless line" case.

Transformed impedances at 220, 230, and 240 MHz are found in a like manner and together form the antenna terminal impedance curve Z_{AA}.

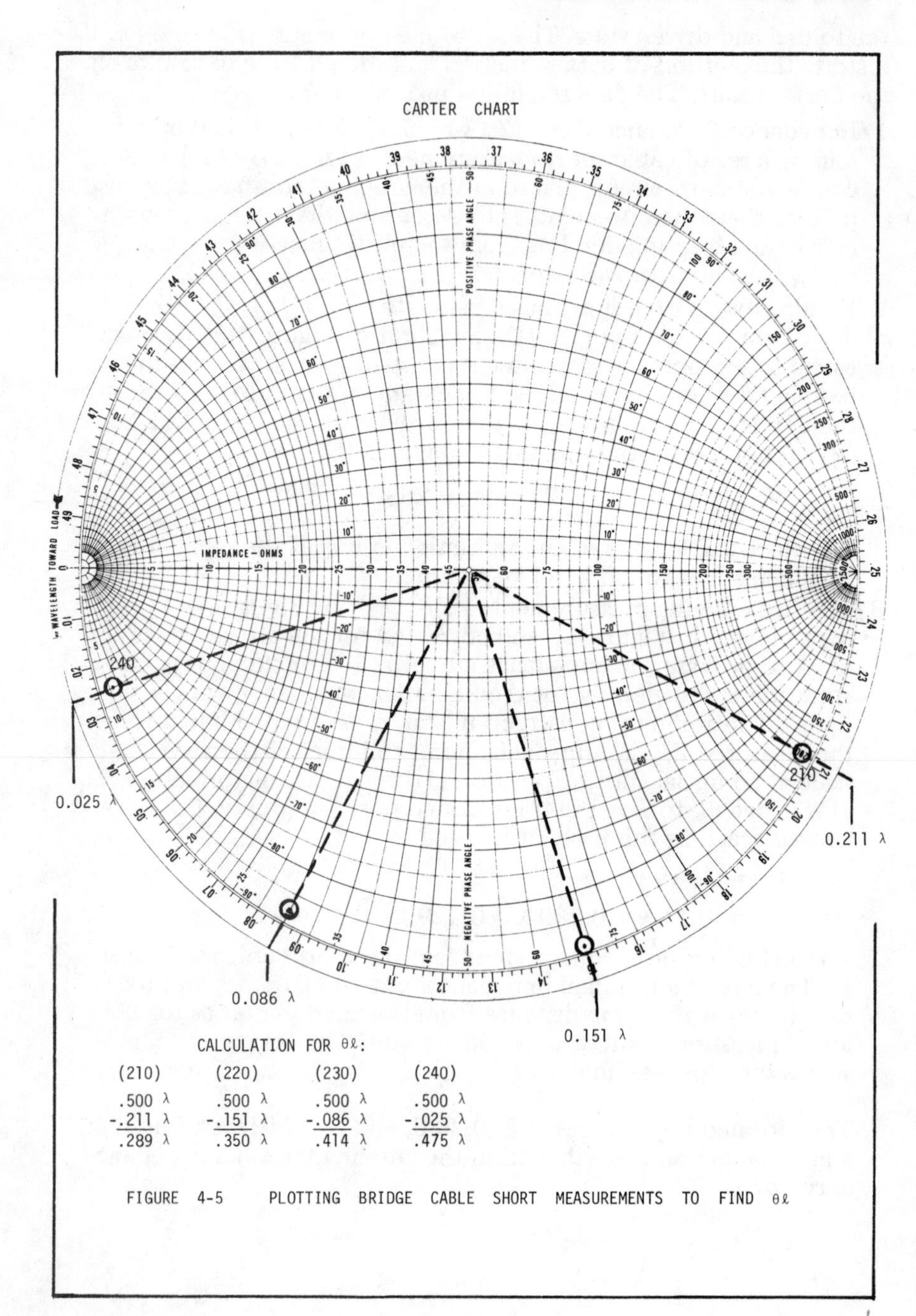

FIGURE 4-5 PLOTTING BRIDGE CABLE SHORT MEASUREMENTS TO FIND θℓ

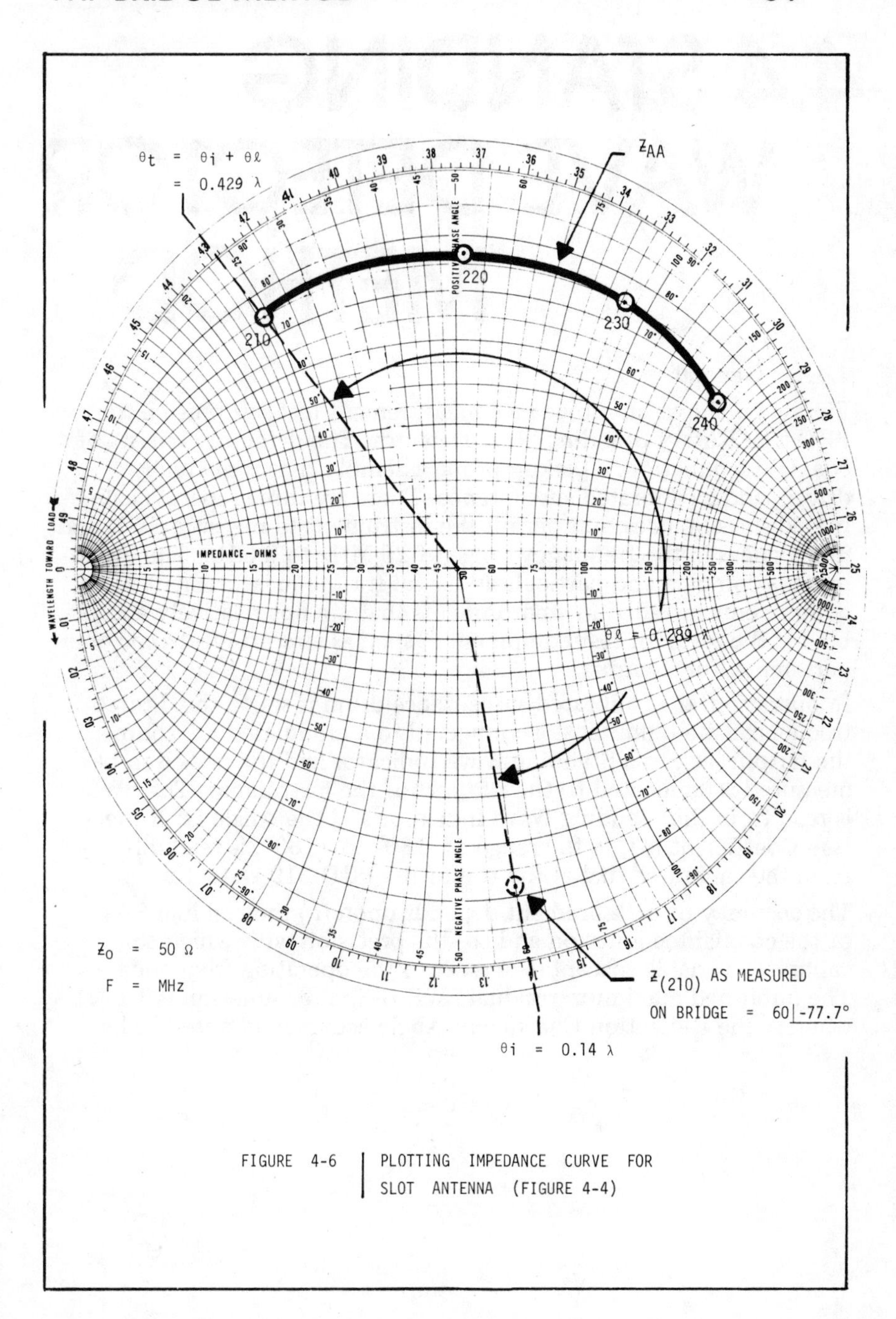

FIGURE 4-6 | PLOTTING IMPEDANCE CURVE FOR SLOT ANTENNA (FIGURE 4-4)

4.4 STANDING WAVE DETECTOR METHOD

4.4.1 General

Figure 4-7 shows an impedance measurement test setup using the PRD 219 Rotary Standing Wave Detector. The detector consists of a coaxial tee junction, a pickup probe assembly that is manually driven, and a calibrated susceptance. There are three separate calibrated susceptances that permit the instrument to measure SWR and impedances in the frequency range of 20 MHz to 2 GHz. The instrument is compact and particularly useful for the measurement of r.f. components installed in environments that are not relatively accessible for the setup of test equipment — such as aircraft electronic compartments, for example.

In operation, the calibrated susceptance drum is set to the desired frequency, the input jack is connected to a modulated r.f. source, the output jack to a Standing Wave Indicator and the device to be measured is connected to the r.f. port of the instrument. The SWR is read from the Standing Wave Indicator and the Angle Of Reflection Coefficient (ϕ), in the range of -180° to +180°, is read directly from the calibrated dial at the top of the PRD 219 Detector.

The accuracy of measurement depends upon (a) the residual SWR of the coaxial tee junction and (b) the precision with which the calibrated variable susceptance is set to the operating frequency. The published maximum residual SWR of the tee junction is 1.04:1 whereas the Reflection Coefficient Angle accuracy is stated to be ± 5°.

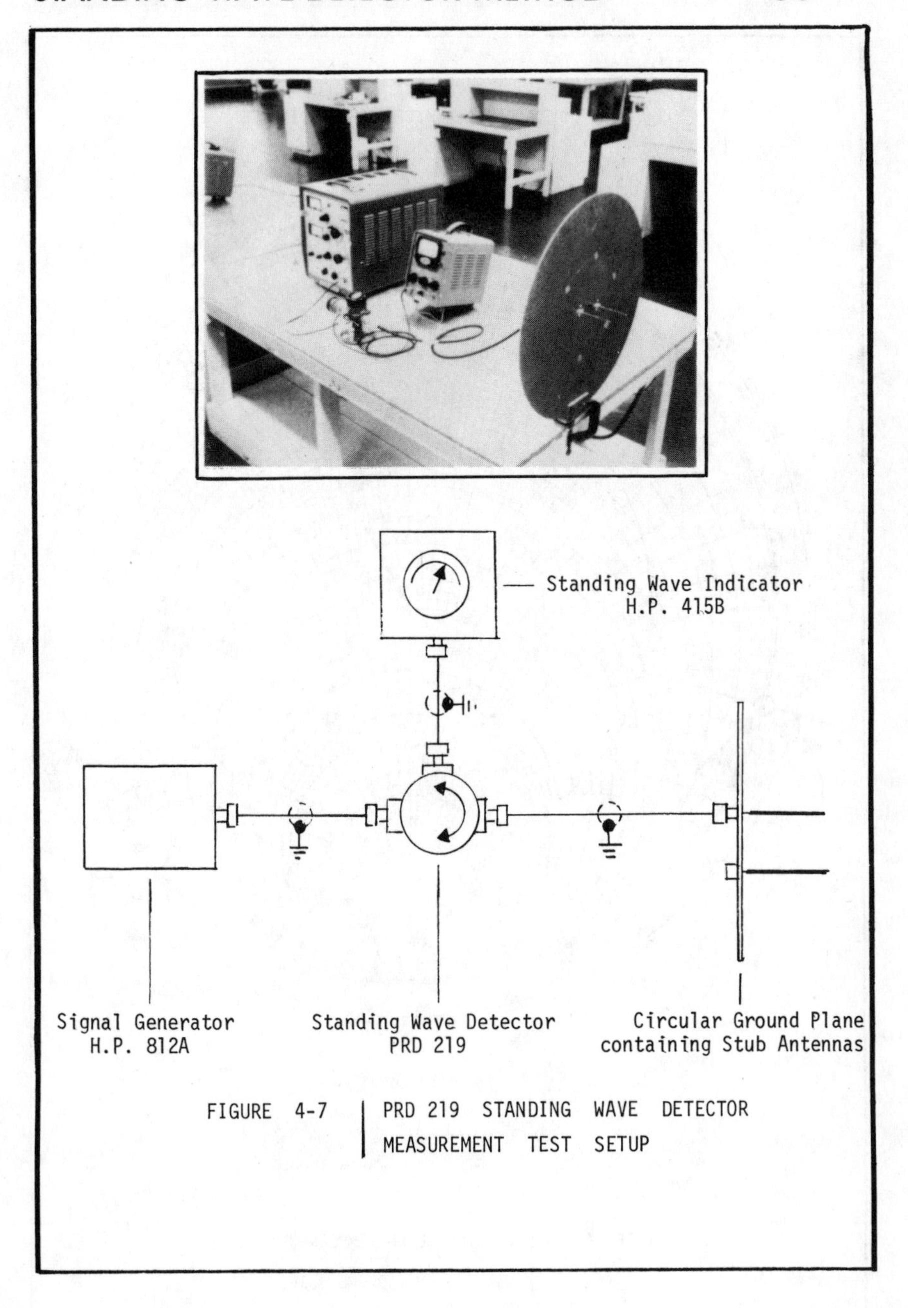

FIGURE 4-7 | PRD 219 STANDING WAVE DETECTOR MEASUREMENT TEST SETUP

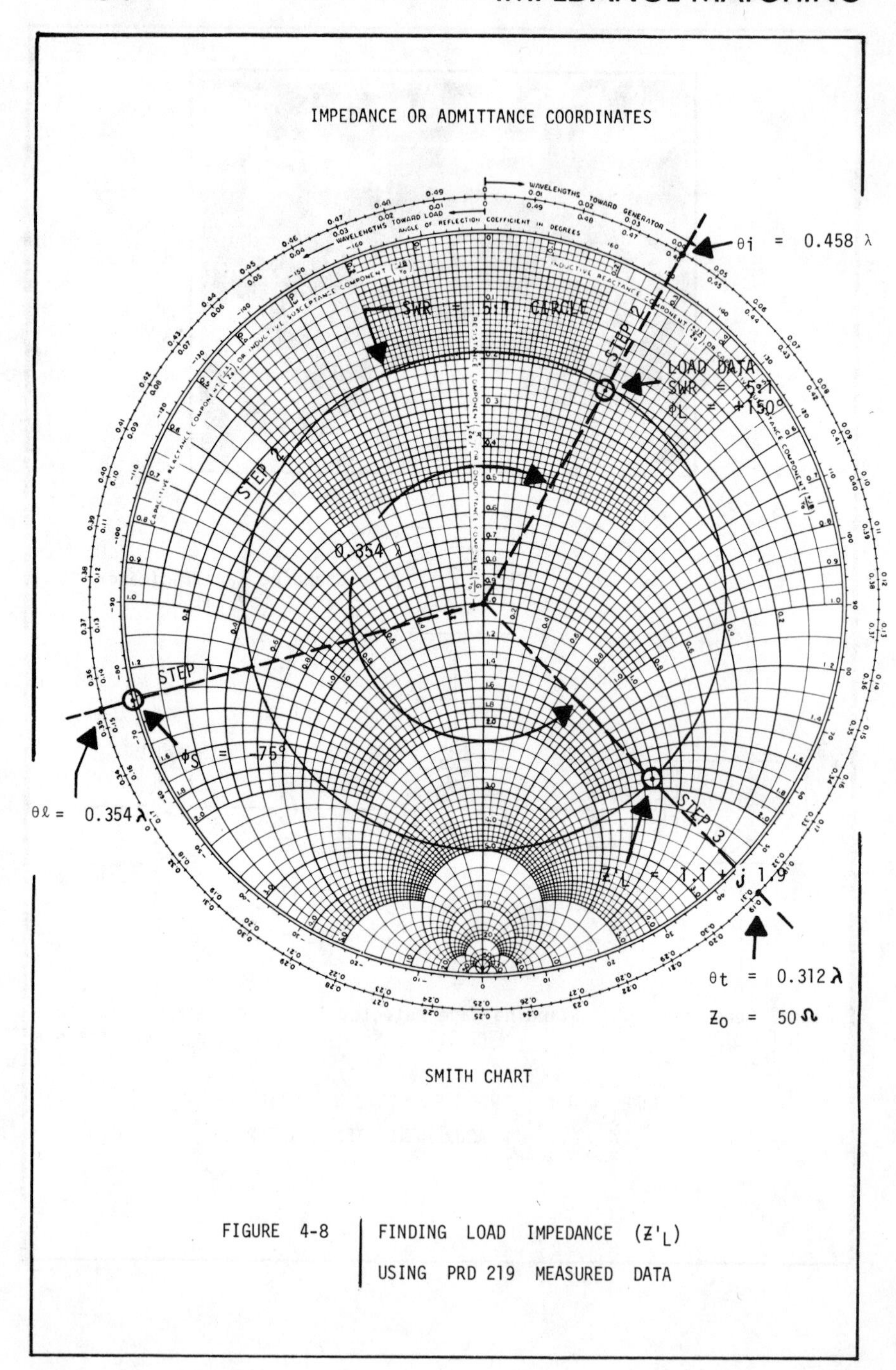

FIGURE 4-8 | FINDING LOAD IMPEDANCE (Z'_L) USING PRD 219 MEASURED DATA

4.4.2 Equipment Calibration

The r.f. port of the PRD 219 Detector is a coaxial line designed to have a characteristic impedance of 50 ohms. The calibration of the equipment may be verified by obtaining calibrated open and short-circuited readings of this port and calculating the Zo using Equation (1-10). The SWR accuracy may be verified by the measurement of standard mis-matches.

In addition, the calibration of the phase angle dial may be verified to within an accuracy of 2° by the short circuit measurement. A perfect short will reflect an infinite SWR and a phase angle reading of ±180°

The maximum error in SWR will occur for a pure resistive load and will be about equal to the percent error in the setting of the frequency of operation. The maximum error in Angle of Reflection Coefficient occurs for the maximum reactive value possible for a given load SWR and will be about one degree for each percent of error in frequency setting.

4.4.3 Data Reduction

The data obtained from the PRD 219 test setup is in the form of SWR and Angle of Reflection Coefficient (ϕ) and may, therefore, be plotted upon the Smith Chart. The plotted values are normalized to 50 ohms. The measured data consists of (a) Angle of Reflection Coefficient (ϕs) of a short circuit placed at the desired Impedance Reference Plane, (b) the load SWR and (c) the Load Angle of Reflection Coefficient (ϕ_L).

The method of data reduction is best described by example. Let it be assumed that it is desired to plot the following measured data:

SWR = 5:1
ϕs = -75°
ϕ_L = +150°

This data is plotted upon the Smith Chart of Figure 4-8 in the following steps:

1/The short measurement, ϕs = -75°, is plotted upon the chart as Point A. A radial line is constructed from the chart center through

Point A to intersect the Wavelengths Towards Generator Scale. The point of intersection defines the electrical length ($\theta \ell$) of the transmission line and, in this case, is equal to 0.354 λ.

2/A SWR = 5:1 circle is next identified in the figure. A radial is then constructed from the chart center to intersect $\phi_L = +150°$ and the Wavelengths Towards Load Scale. The point of intersection of the radial line and the SWR = 5:1 Circle defines the load impedance as measured at the plane of the PRD 219 Detector. The point of intersection of the radial line and the Wavelengths Towards Load Scale defines the initial starting position (θi) for line transformation which in this case is 0.458 λ.

3/Line transformation is accomplished by adding θi and $\theta \ell$ to give the transformed position (θt) on the Wavelengths Towards Load Scale as:

$$\begin{aligned} \theta t &= \theta i + \theta \ell \\ &= 0.458\,\lambda + 0.354\,\lambda \\ &= 0.812\,\lambda \text{ or } 0.312\,\lambda \end{aligned}$$

A radial line is constructed from the chart center through θt = 0.312 λ. The point of intersection of this line and the SWR = 5:1 circle establishes the normalized load impedance (Z'_L) to be 1.1 + j 1.9. The actual value of the load impedance (Z_L) is:

$$\begin{aligned} Z_L &= Z'_L \cdot 50 \\ &= 55 + j\,95\ \Omega \end{aligned}$$

The above procedure assumes the "lossless line" case. Had there been line loss, the load impedance (Z'_L) would require further correction using the method described in paragraph 3.2.2 of Section III.

4.5 SLOTTED LINE METHOD

4.5.1 General

The slotted line is another instrument that measures impedance utilizing the technique of measuring the Standing Wave Ratio (SWR) that exists on a transmission line as a result of a load mis-match. Slotted lines are available in both coaxial and waveguide types. Coaxial lines generally cover the frequency range from 500 MHz to 18 GHz, whereas waveguide types are utilized in the frequency range of 2 GHz to 40 GHz.

Figure 4-9 shows a typical test setup. Both coaxial and waveguide lines are shown to illustrate the two principal types. In practice, only one slotted line per test setup is generally required. The Slotted Line contains an adjustable probe that may be extended into the slotted section to sample the r.f. voltage. This voltage is rectified by a detector contained within the probe mount, amplified, and displayed on a SWR Indicator. In operation the probe is moved along the line to measure the maximum and minimum of the voltage Standing Wave pattern. The probe is "loosely" coupled to the line for this measurement, in order to cause minimum distortion of the r.f. field within the guide.

The accuracy of slotted line measurements is limited primarily by the residual SWR of the line. The residual SWR is that SWR measured when the line is terminated in a perfect match. Residual SWR for coaxial lines varies from 1.02:1 to 1.06:1 whereas waveguide lines reflect a residual SWR of 1.01:1. Other considerations, such as probe penetration, harmonics, and detector "square law" response affect the measurement accuracy, however, these factors are controllable by proper operational procedures.

4.5.2 Equipment Calibration

Calibration of Slotted Line test setups include (a) shorted line measurements, (b) detector "square law" response, and (c) calibrated mis-match measurements.

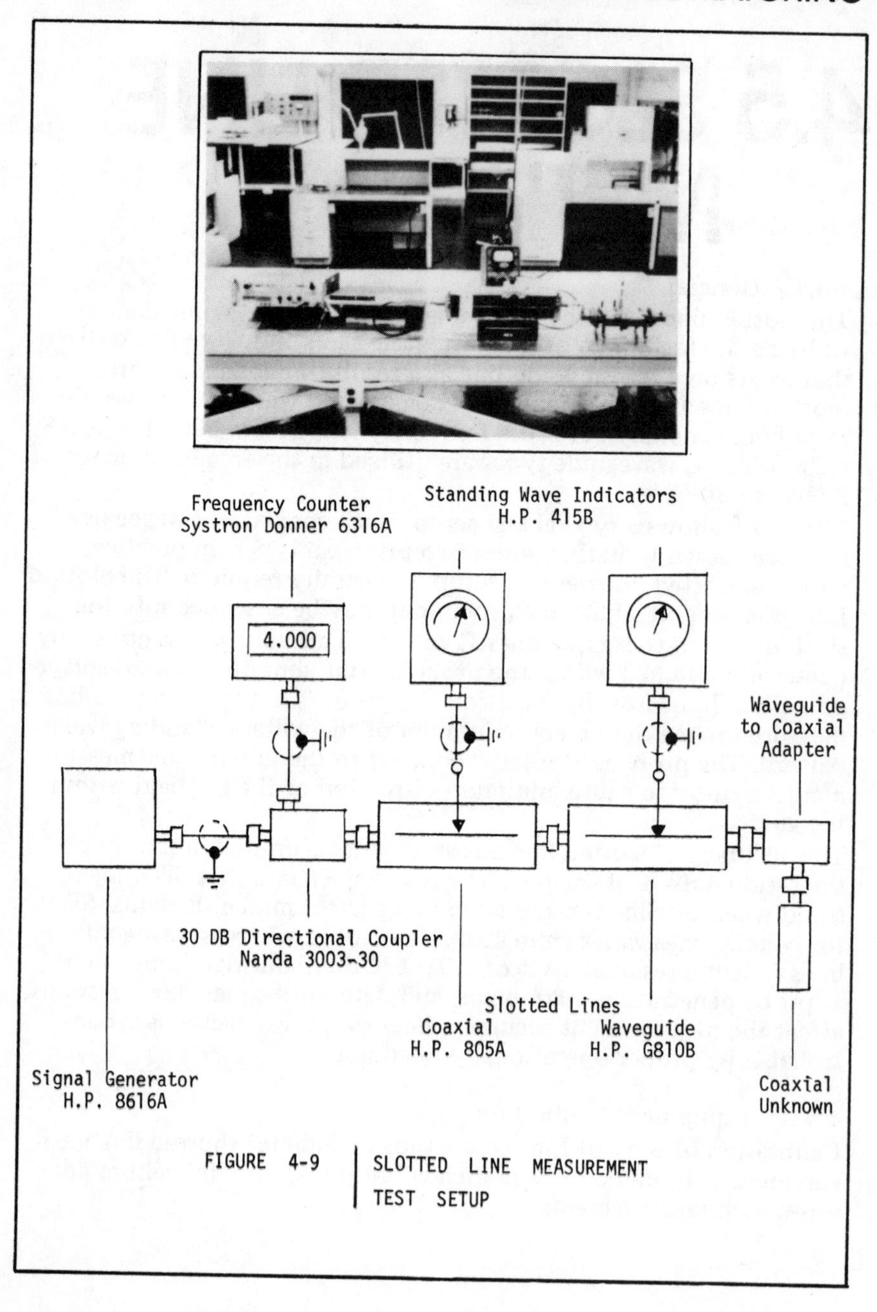

FIGURE 4-9 | SLOTTED LINE MEASUREMENT TEST SETUP

The shorted line measurement consists of placing a short at the Impedance Reference Plane and measuring the distance between two successive minima. This measurement multiplied by two defines the guide wavelength (λg) and hence provides a means of verification of the frequency of operation and probe tuning. In an air-filled coaxial Slotted Line, λg for all practical purposes is equal to the free space wavelength (λ_0) as defined in Equation (1-1) whereas for the air-filled rectangular waveguide, λg is given by Equation (1-13). If there is an excessive degree of r.f. harmonics emitted from the signal source, the probe may very easily be improperly tuned to the second harmonic instead of the fundamental. Comparison of calculated and measured values of λg thus provides a means of frequency calibration. The short measurement also describes the presence of FM. Sharp deep nulls indicate FM not excessive whereas broad nulls is a clear indication of excessive FM from the signal source. Excessive FM results in significant errors in SWR measurements.

The "square law" response of the probe detector may be verified by decreasing the source power output in discrete increments (usually 3 dB increments) and noting the response on the SWR Indicator. If the SWR Indicator response closely follows that of the signal source incremental power decrease, the detector is thus operating in the "square law" region.

The measurement accuracy may be verified by the measurement of standard mis-matches that are calibrated for specific SWR values. The standard mis-matches should have SWR values comparable to that of the r.f. component to be measured.

4.5.3 Data Reduction

The measurement of impedance, using a Slotted Line, requires (a) a measurement of two successive voltage minima for the shorted line case and (b) a measurement of the SWR of the load. The measurement data, at a single frequency, shown in Figure 4-10 is selected to illustrate data reduction. The technique is as follows:

1/The antenna terminals are shorted to produce the Standing Wave pattern shown in the figure. Two successive voltage nulls are noted — one towards the load (N_L) and the other towards the generator (N_G) The guide wavelength (λg) is found as:

$$\lambda g = (N_L - N_G) \cdot 2 \quad (4\text{-}2)$$
$$= (27.94 - 5.23) \cdot 2$$
$$= 45.42 \text{ cm} = 17.88 \text{ inches}$$

The calculated free space wavelength (λ_o) from Equation (1-1) for a coaxial air-filled line is

$$\lambda_o = \frac{(300)(39.4)}{\text{frequency (MHz)}}$$
$$= \frac{(300)(39.4)}{660}$$
$$= 17.92 \text{ inches}$$

Close correlation of the above measured wavelength values provides a verification of frequency setting and probe tuning.

2/The SWR of the antenna is next measured and an antenna null position (N_A) that lies between N_G and N_L is determined. In this case these values are:

$$\text{SWR} = 3.5{:}1$$
$$N_A = 14.82 \text{ cm}$$

3/N_L, N_G, N_A, and SWR comprises the measured data needed to form a Smith Chart plot. Figure 4-11 is a plot of the measurements given in Figure 4-10. In Figure 4-11 a SWR circle of 3.5:1 is first constructed. The impedance to be determined must lie somewhere upon this circle. The angular position ($\theta \ell$), in wavelengths, needed to establish the exact location of impedance is derived from the N_L, N_G, and N_A data.

4/$\theta \ell$ determination — $\theta \ell$ may be plotted either towards the generator or towards the load on the wavelengths scales. Either plot

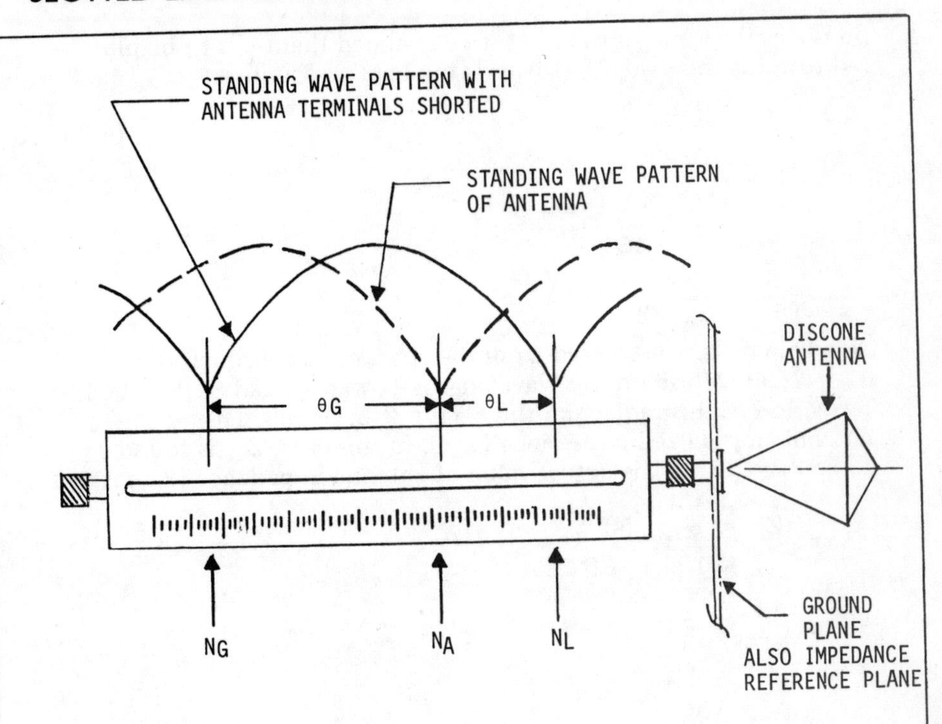

MEASURED DATA:

FREQUENCY (MHZ)	SWR	N_G (CM)	N_A (CM)	N_L (CM)
660	3.5:1	5.23	14.82	27.94

WHERE:

N_G = NULL TOWARDS GENERATOR

N_L = NULL TOWARDS LOAD

N_A = ANTENNA NULL

FIGURE 4-10 | SLOTTED-LINE MEASUREMENT DATA FOR A DISCONE ANTENNA

produces the same results. Let it be assumed that $\theta \ell$ is to be plotted towards the load. $\theta \ell$ (towards the load) is found as:

$$\theta \ell = \frac{N_L - N_A}{\lambda g} \qquad (4\text{-}3)$$

$$= \frac{27.94 - 14.82}{45.42}$$

$$= 0.289\ \lambda$$

A radial line is constructed from the chart center through the $\theta \ell = .289$ position on the Wavelengths Towards Load Scale. The intersection of this radial and the SWR = 3:5:1 circle defines the antenna normalized impedance (Z'_A). In this case, Z'_A is found to be 2.1 + j 1.6. The actual value of antenna impedance (Z'_A) is:

$$Z_A = Z'_A \cdot 50$$
$$= 105 + j\ 80\ \Omega$$

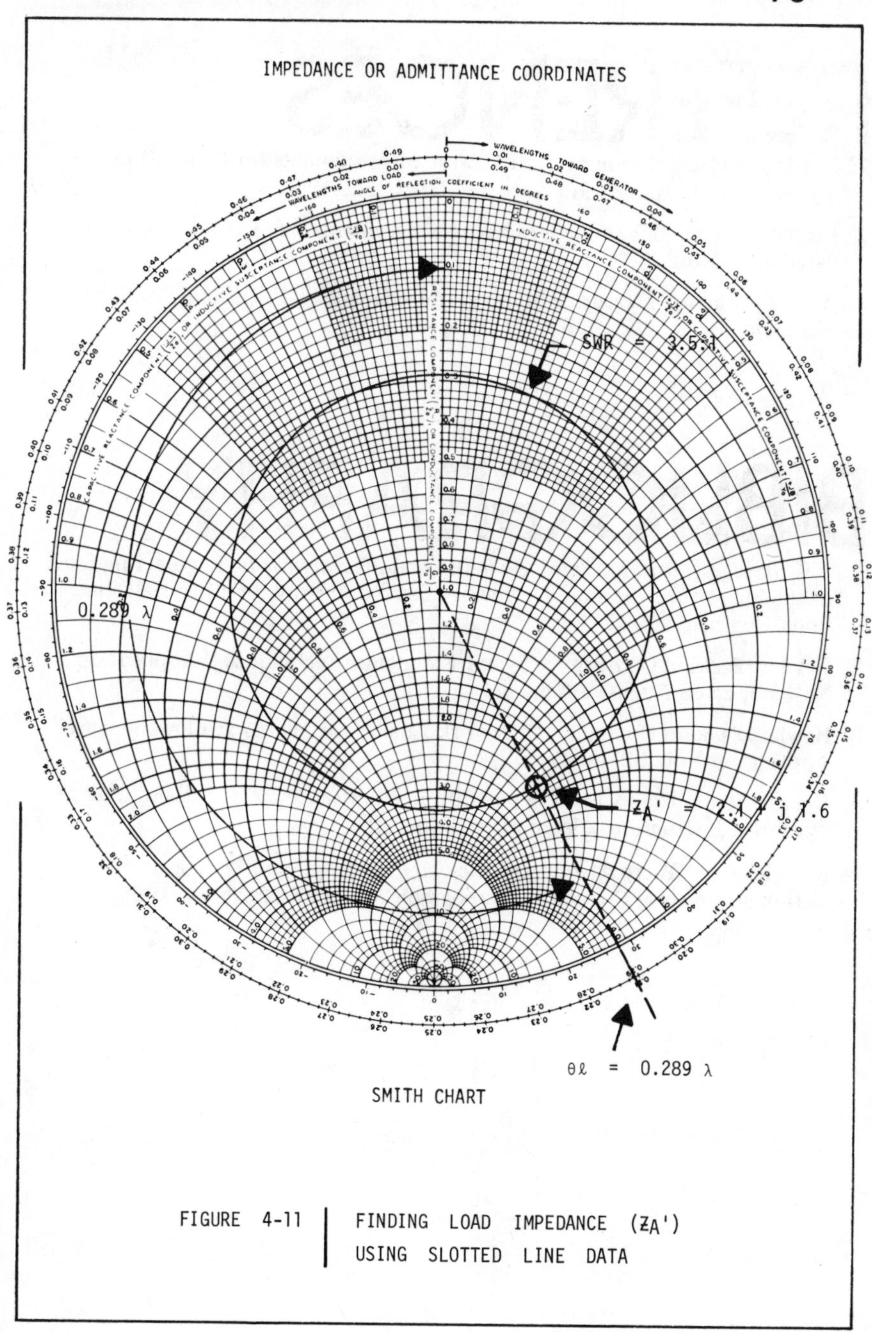

FIGURE 4-11 | FINDING LOAD IMPEDANCE (Z_A') USING SLOTTED LINE DATA

REFERENCES

1. Hewlett-Packard Company, "Microwave Measurements for Calibration Laboratories," Application Note 38, dated June 1960.
2. Pat Tucciarone, "Making Microwave Measurements," Electronic Industries, dated June 1962.
3. W.A. Weissman, "Measuring Microwave Impedance," *Electronic Products Magazine*, dated May 1966.

BIBLIOGRAPHY

1. L.K. Irving, Editor, *Microwave Theory and Measurements*, Engineering Staff of the Microwave Division, Hewlett-Packard Company, Prentice-Hall, Inc. Book Company, 1962.
2. Hewlett-Packard Company, "Swept Frequency Techniques," Application Note 65, dated August 1965.
3. Hewlett-Packard Company, "Network Analysis at Microwave Frequencies," Application Note 92, no date.
4. General Radio Company, "Type 1602-B Admittance Meter," Handbook of Operating Instructions, dated May 1965.
5. R.W. Beatty, "Microwave Impedance Measurements and Standards," National Bureau of Standards, Monograph 82, dated 12 August 1965.

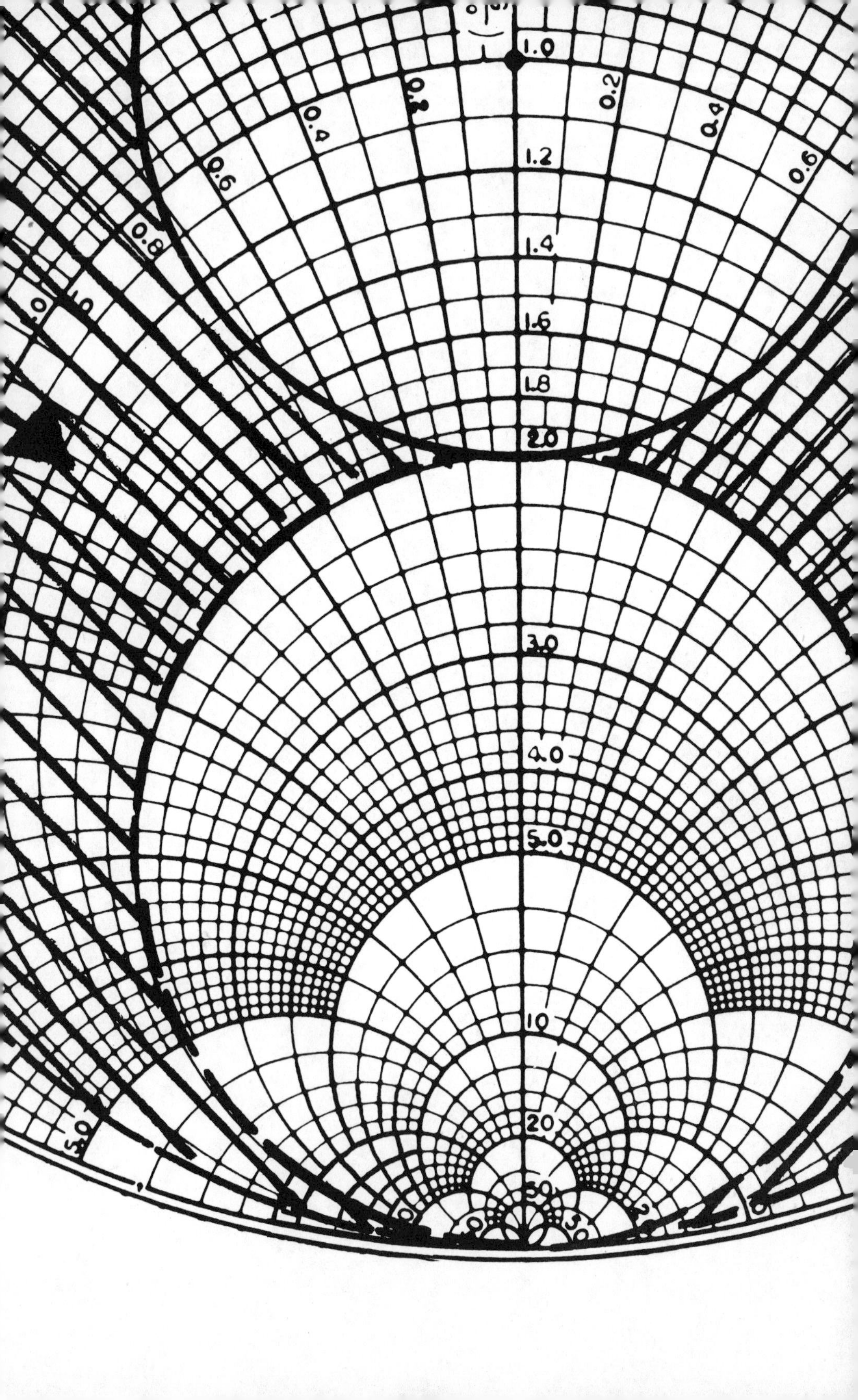
1.0
0.2
0.4
0.6
0.8
1.2
1.4
1.6
1.8
2.0
3.0
4.0
5.0
10
20
50

5 NARROW BAND MATCHING

5.1 INTRODUCTION

Single frequency or narrow band impedance matching is quite easily accomplished with a minimum of development effort. Moderate "Q" r.f. structures may often be matched with a single fixed-tuned network whereas high "Q" structures would require, at the most, a two-element network. A two-element network would provide a perfect or near perfect match at a single frequency.

Networks applicable to narrow band matching are discussed in this section. Such networks include (a) single-element series and shunt types, (b) two-element types, and (c) simple quarterwave transformers that use both "lumped" and "distributed" parameters. Smith Chart illustrations are shown that describe chart areas within which impedances are susceptible to compensation using series or shunt networks. Such illustrations as these facilitate network selection.

Practical application of narrow band matching methods are related to stub tuning devices that are used in scale model antenna radiation pattern measurements.

5.2 DEFINITION OF BANDWITH

The bandwidth of an r.f. component is that frequency range over which the component operates within a specified performance requirement. The performance requirement for impedance bandwidth is specified in terms of SWR. Percent bandwidth (%BW) is expressed as:

$$\%BW = 2\left(\frac{f_u - f_L}{f_u + f_L}\right) \cdot 100 \tag{5-1}$$

Where: f_u = upper frequency limit

f_L = lower frequency limit

A VHF aircraft antenna, for example, that operates within the 118 MHz to 136 MHz frequency band is said to have a 14.2% bandwidth in accordance with the above definition of bandwidth.

The term "octave" is another expression that is commonly used in r.f. work to describe bandwidth. To illustrate, a component that operates from 2000 MHz to 4000 MHz is said to operate an octave in bandwidth. Had the upper frequency been 8000 MHz, the same component could be described as a two-octave device.

Other frequently used terms relating to bandwidth and the classification of r.f. components are (a) narrow band, (b) moderate band, and (c)broad band. Component classifications on the basis of these terms are somewhat vague; however, typical classification of bandwidth limits that are generally acceptable throughout the industry are:

Bandwidth Term	*General Limits*
Narrow Band	Less than 10%
Moderate Band	Between 10% and 50%
Broad Band	Greater than 50%

5.3 SERIES AND SHUNT NETWORKS

5.3.1 Matching Analysis Aids

Figures 5-1 and 5-2 are graphical design aids that may be used to determine network configurations for series and shunt elements to compensate a given load impedance. The 2:1 SWR circle that is shown in the figures is arbitrarily selected for illustration only. Similar constructions are quite easily defined for other SWR value circles. The shaded chart areas in the figures define those areas in which impedances so located are susceptible to matching using series and shunt elements. The application of these graphical aids will be described for single and two-element networks such as those that are shown in Figure 5-3.

5.3.2 Single-Element Networks

A single-element, either in series or shunt with a load impedance, forms the most simple type of matching network. The network adds a reactance ($\pm jX$) or susceptance ($\pm jB$) component that tends to cancel that respective component of the complex load impedance. The compensation effectiveness of a single-element network is dependent upon the chart location of the initial impedance curve. Consider the load impedance curve, Z'_{AA}, that is shown in Figure 5-4. The load impedances normalized to 50 ohms are:

f (MHz)	*Load Impedance* (Z'_{AA})
30.0	$0.5 + j\,1.0$
31.0	$0.8 + j\,1.6$
32.0	$2.0 + j\,2.5$

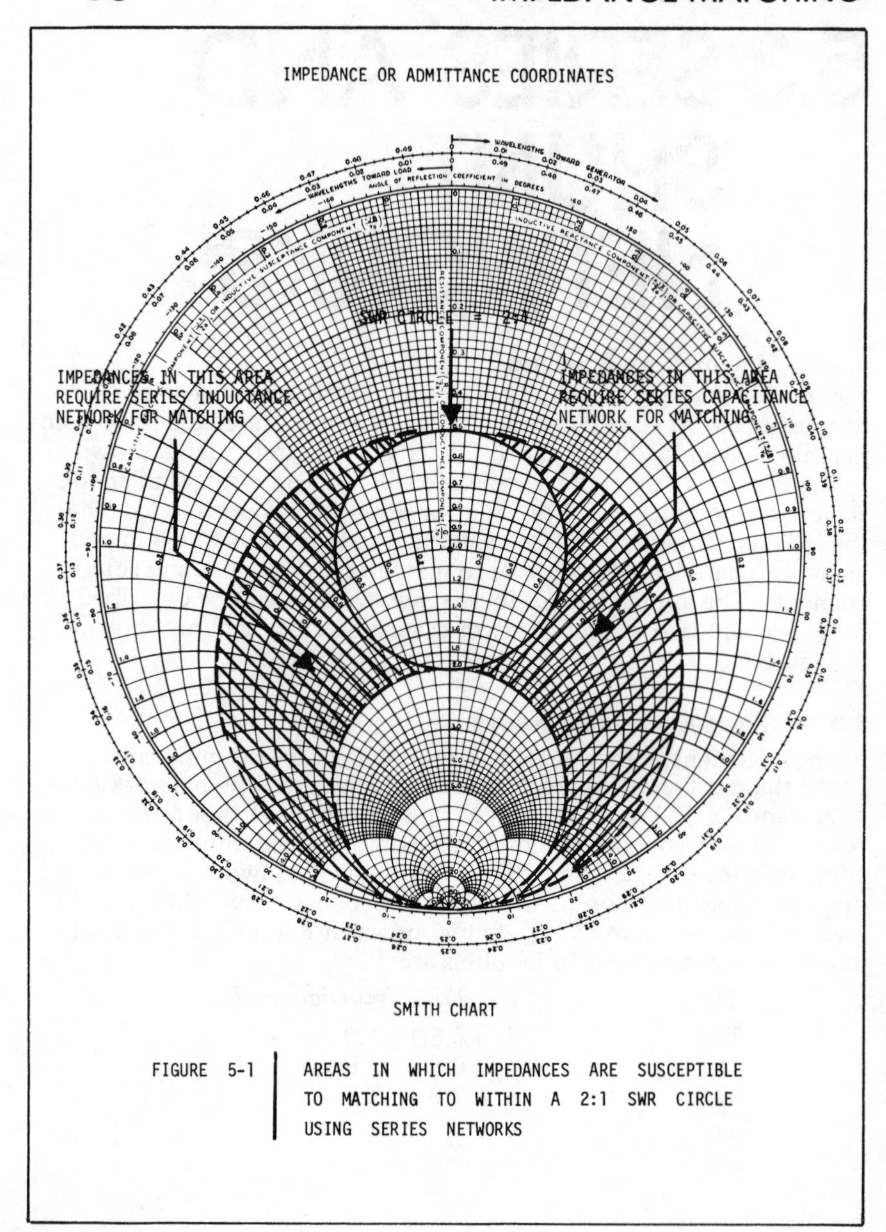

FIGURE 5-1 AREAS IN WHICH IMPEDANCES ARE SUSCEPTIBLE TO MATCHING TO WITHIN A 2:1 SWR CIRCLE USING SERIES NETWORKS

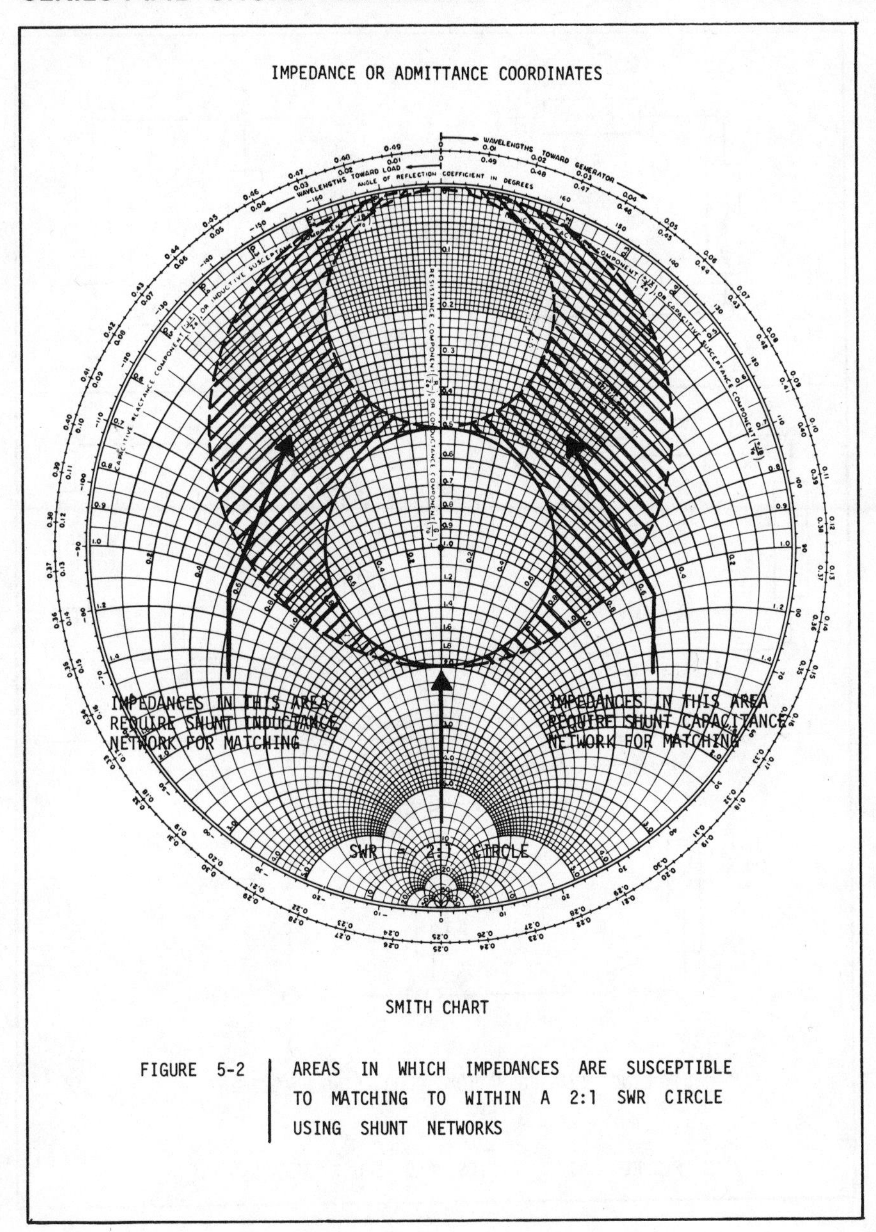

FIGURE 5-2 AREAS IN WHICH IMPEDANCES ARE SUSCEPTIBLE TO MATCHING TO WITHIN A 2:1 SWR CIRCLE USING SHUNT NETWORKS

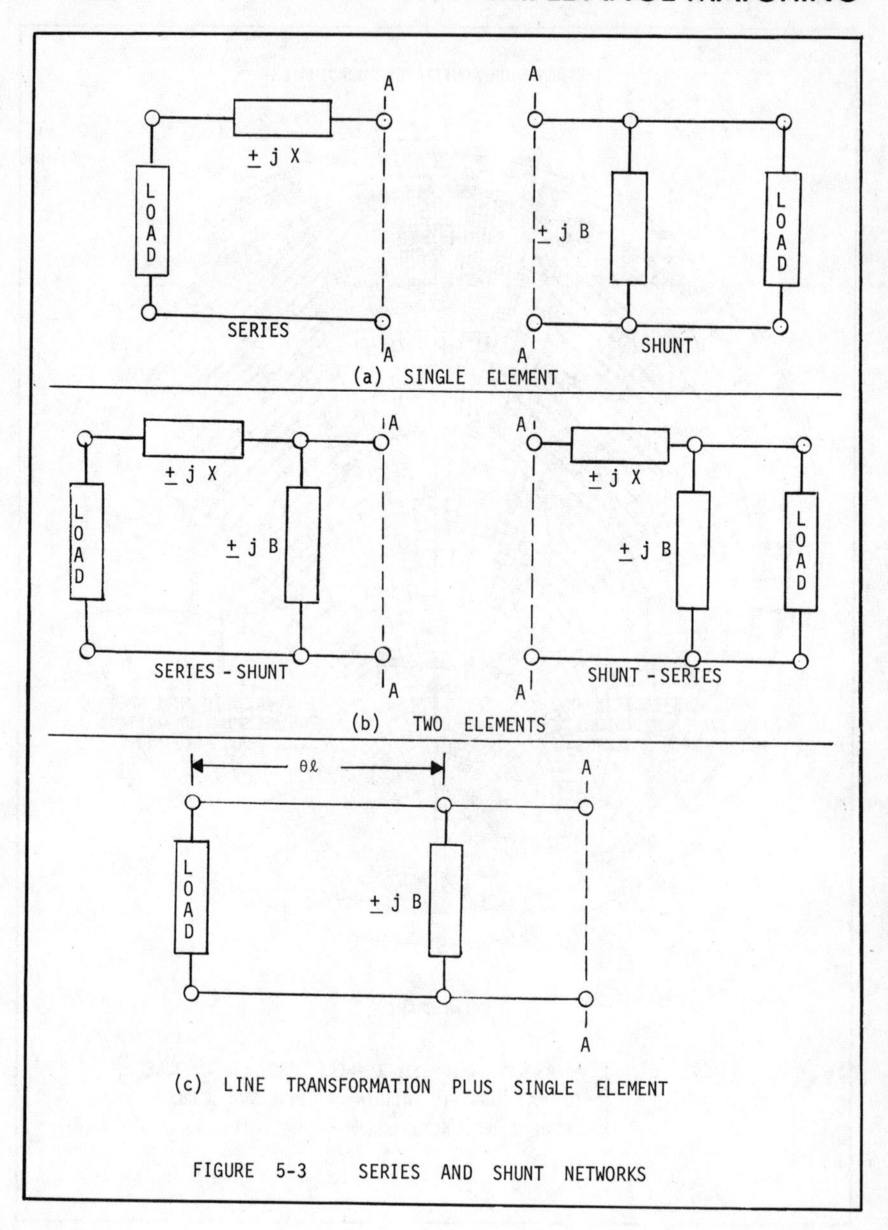

FIGURE 5-3 SERIES AND SHUNT NETWORKS

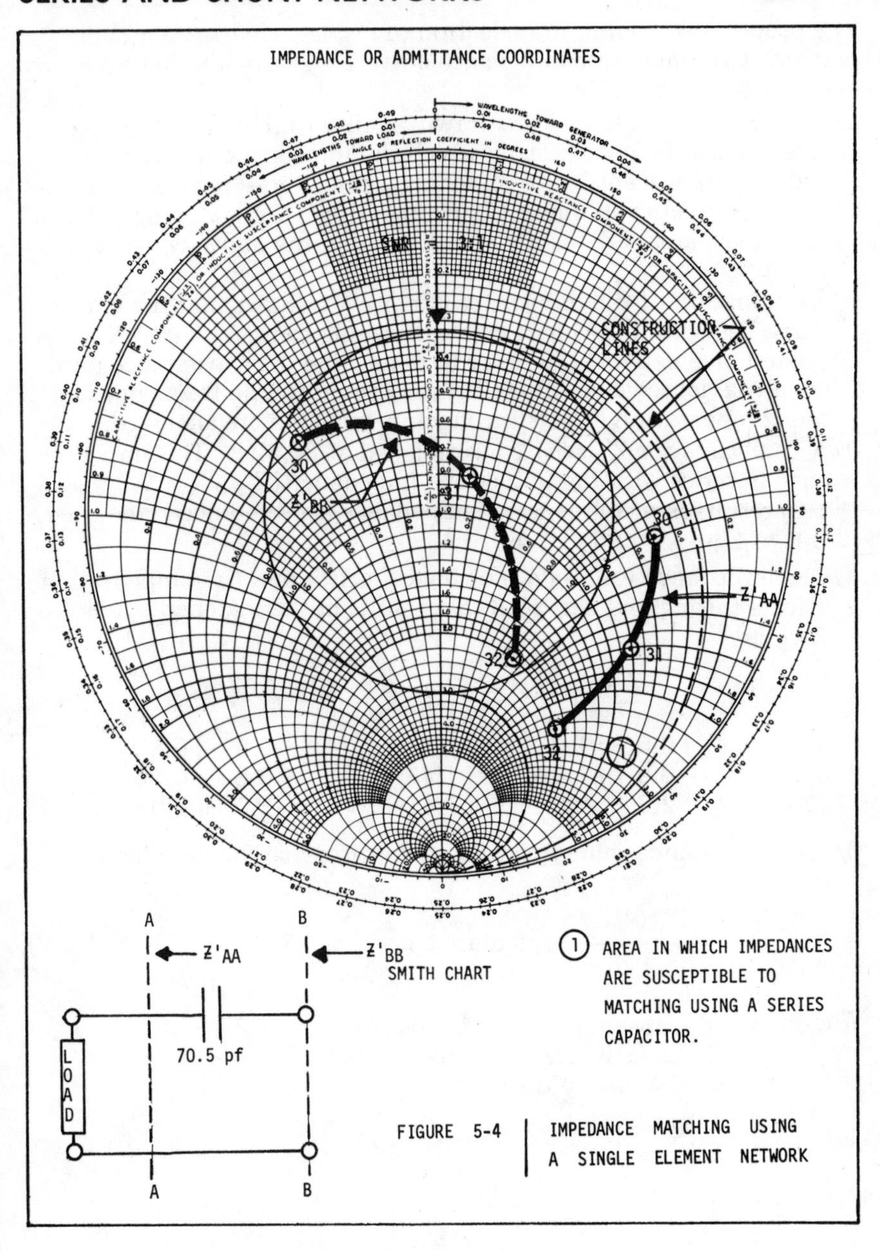

FIGURE 5-4 | IMPEDANCE MATCHING USING A SINGLE ELEMENT NETWORK

The process involved in the selection of a single-element matching network to compensate impedance curve Z'_{AA} to within an SWR circle of 3:1 is as follows:

1/Construction dash lines are drawn, similar to that of Figure 5-2, to establish the chart area in which the impedances must lie in order to be susceptible to matching using a series network. As one achieves practice in Smith Chart operations, the matching susceptible chart areas are readily visualized, and the above construction is not necessary.

2/Examination of the impedance curve Z'_{AA} shows that the entire curve lies within the chart area susceptible to matching by a series capacitance network. In order to determine the capacitance needed, a trial impedance value (Z'_c) is assigned to one frequency and the compensation effects throughout the band are analyzed.

3/A trial value for Z'_c of -j 1.5 is assigned at 30 MHz. This impedance in series with the initial load impedance (Z'_{AA30} = 0.5 + j 1.0) results in a compensated impedance at Reference Plane B-B of Z'_{BB30} = 0.5 - j 0.5.

4/The impedance values (Z'_c) of the capacitor and its compensation effects throughout the band that results in the matched curve Z'_{BB} are:

f (MHz)	Z'_{AA}	+	Z'_c	=	Z'_{BB}
30.0	0.5 + j 1.0		- j 1.5		0.5 - j 0.5
31.0	0.8 + j 1.6		- j 1.45		0.8 + j 0.15
32.0	2.0 + j 2.5		- j 1.41		2.0 + j 1.09

5/The capacitance value (pf) of the capacitor may be calculated from:

$$C = \frac{10^{12}}{(2\pi f)(X_c)} \quad \text{(picofarads)} \tag{5-2}$$

Where: C = capacitance (picofarads)
X_c = capacitance reactance (ohms)
f = frequency (hertz)

The capacitance reactance (X_c) at 30 MHz is:

$$X_{c30} = Z'_{c30} \cdot 50$$
$$= -j\,1.5 \cdot 50 = 75 \text{ ohms}$$

and from Equation (5-2):

$$C = \frac{10^{12}}{(6.28)\,(30 \times 10^{6})\,(75)} = 70.5 \text{ pf}$$

Thus the chart location of the initial impedance curve, Z'_{AA}, was favorable such that compensation over the frequency band was accomplished by a single-element capacitor network.

5.3.3 Two-Element Networks

Two-element matching networks are less dependent upon the chart location of the initial impedance curve and are more flexible in application as compared to the single-element network. Also, two -element networks provide greater bandwidths when properly selected. In Section VII, a theory of broadband impedance matching[1] is presented that describes the method of selecting the first two elements, in a four-element configuration, that will provide the maximum impedance bandwidth. It will suffice here in the present section, however, to describe only the manner in which a two-element network compensates a given impedance curve.

In Figure 5-5 is shown the impedance curve, Z'_{AA}, of a high "Q" structure whose normalized values over the operating frequency band are:

f (MHz)	Z'_{AA}
95.0	0.6 - j 4.0
97.5	0.5 - j 3.7
100.0	0.4 - j 3.5

Let it be desired to compensate impedance curve Z'_{AA} to within an SWR = 2:1 circle using a two-element network. For the purpose of this example, let the network elements be arranged in a series-shunt configuration. The process of impedance matching follows:

1/A series-shunt network configuration, in this example, requires that the first element be a series inductance and the second element a shunt inductance. The matching effect desired of the first element is to move the initial impedance curve to within the chart area where impedances are susceptible to matching using a shunt network. Construction dash lines are therefore established in Figure 5-5, similar to that of Figures 5-1 and 5-2 to establish the matching susceptibility chart areas ① and ②.

2/A trial impedance value for the first element of Z'_{L1} = + j 3.1 is assigned at 95 MHz. This value in series with the load impedance (Z'_{AA95} = 0.6 - j 4.0) provides an impedance at Reference Plane B-B of Z'_{BB95} = 0.6 - j 0.9.

Z'_{BB95} falls just within the matching susceptibility chart area 1. The compensating effect of Z'_{L1} on impedance curve Z'_{AA} throughout the frequency band is:

f (MHz)	Z'_{AA}	+	Z'_{L1}	=	Z'_{BB}
95.0	0.6 - j 4.0		+ j 3.1		0.6 - j 0.9
97.5	0.5 - j 3.7		+ j 3.18		0.5 - j 0.52
100.0	0.4 - j 3.5		+ j 3.25		0.4 - j 0.25

The trial value selected for Z'_{L1} is satisfactory since the compensated curve Z'_{BB} falls within the desired chart area 1 that is susceptible to matching using a shunt network.

3/ Admittance curve Y'_{BB} is established diametrically opposite Z'_{BB} as the second element will provide shunt matching. Values of Y'_{BB} as read from Figure 5-5 are:

f (MHz)	Y'_{BB}
95.0	0.52 + j 0.78
97.5	0.95 + j 1.0
100.0	1.80 + j 1.15

A trial value of Y'_{L2} = -j 0.88 is assigned for the admittance of the second element at 95 MHz. This admittance in shunt with Y'_{BB95} (0.52 + j 0.78) gives Y'_{CC95} = 0.52 - j 0.10. Y'_{CC95} falls within

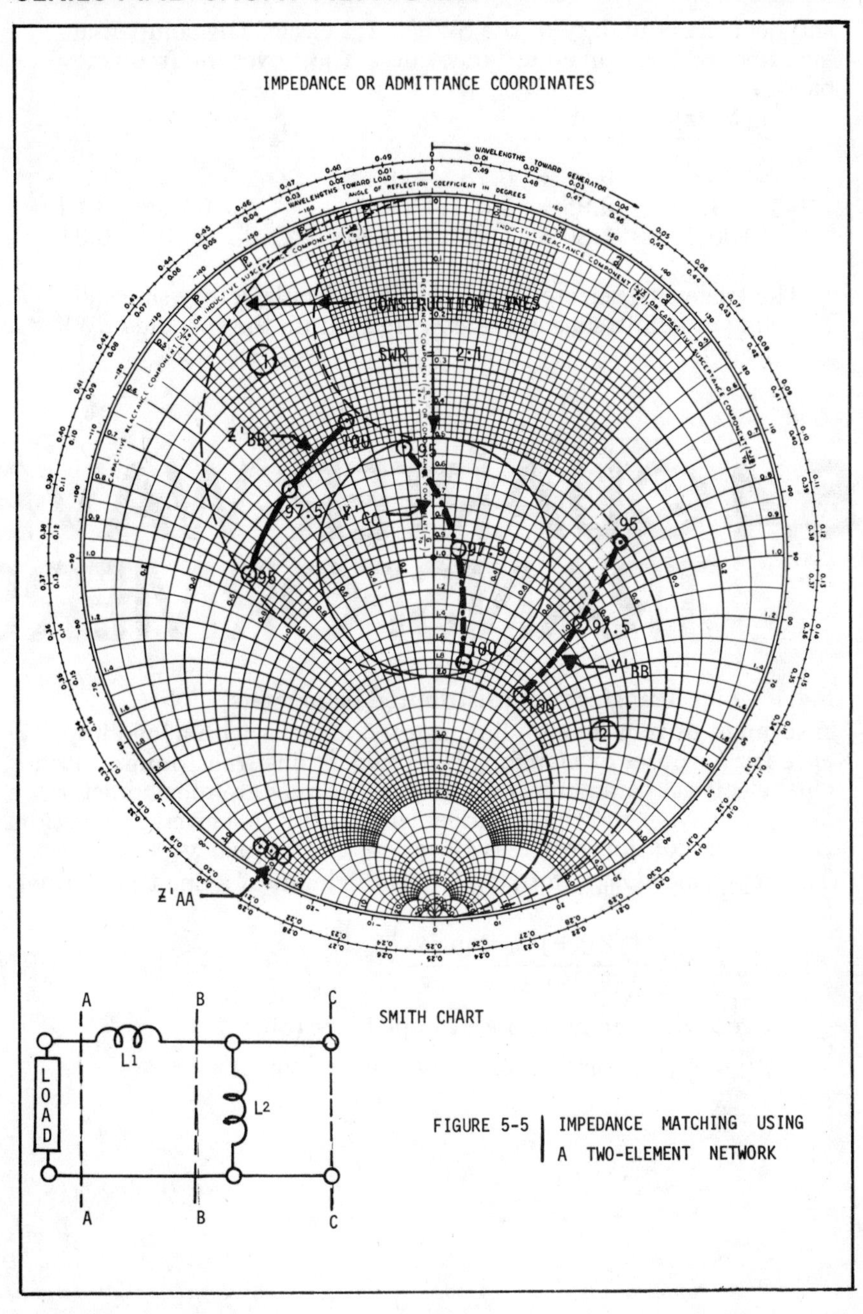

FIGURE 5-5 | IMPEDANCE MATCHING USING A TWO-ELEMENT NETWORK

and near the boundary of the SWR = 2:1 circle. The compensating effect of Y'_{L2} on admittance curve Y'_{BB} over the frequency band is:

f (MHz)	Y'_{BB}	+ Y'_{L2}	= Y'_{CC}
95.0	0.52 + j 0.78	- j 0.88	0.52 - j 0.1
97.5	0.95 + j 1.0	- j 0.86	0.95 + j 0.14
100.0	1.80 + j 1.15	- j 0.84	1.80 + j 0.31

The trial value selected for Y'_{L2} proved satisfactory since all points of admittance curve Y'_{CC} fall within the specified SWR = 2:1 circle.

5.4 QUARTER-WAVE TRANSFORMERS

5.4.1 Line Transformers (distributed parameters)

A segment of transmission line inserted in series with a load impedance for purposes of matching is termed a "line transformer." Principal electrical parameters of a line transformer are the characteristic impedance (Z_0') and the electrical length ($\theta\ell$). In a given application, Z_0' remains constant whereas $\theta\ell$ varies with frequency.

The impedance (Z_{in}) "looking " into the input of a line transformer is:

$$Z_{in} = Z_0' \cdot \left(\frac{Z_L + j Z_0' \tan \theta\ell}{Z_0' + j Z_L \tan \theta\ell} \right) \tag{5-3}$$

Where: Z_{in} = complex input impedance (ohms)

Z_L = complex load impedance (ohms)

Z_0' = transformer characteristic impedance (ohms)

$\theta \ell$ = electrical length of transformer (degrees)

For the frequency at which $\theta \ell = \lambda/4 = 90°$, the transformer is termed a "quarter-wave" transformer and the input impedance is given by:

$$Z_{in} = \frac{(Z_0')^2}{Z_L} \quad (5\text{-}4)$$

A schematic of the transmission line (distributed parameters) quarter-wave transformer is shown in Figure 5-6(a). Quarter-wave transformers are used as matching devices in the higher frequency bands where the physical length of the transformer is not objectionable.
At a single frequency, a resistive load (R_L) may be perfectly matched to the associated transmission line (Z_0) by a quarter-wave transformer whose characteristic impedance (Z_0') is:

$$Z_0' = \sqrt{Z_0 \cdot R_L} \quad (5\text{-}5)$$

Where: Z_0' = transformer characteristic impedance (ohms)

Z_0 = transmission line characteristic impedance (ohms)

R_L = purely resistive load (ohms)

Line transformer usage is not necessarily restricted to narrow band matching applications. Section VI shows their application as broadband matching devices whose Z_0' and $\theta \ell$ are selected by graphical methods.[2]

5.4.2 "PI" and "T" Transformers (Lumped parameters)

Equivalent quarter-wave "PI" and "T" transformers are shown in Figure 5-6(b) and (c). These transformers utilize lumped parameters and are particularly useful in the lower frequency bands where the physical length of a line transformer would be objectionable.
An interesting relationship to the line transformer exists that permits ready determination of the parameter values (Z_L and Z_C) of the "PI" and "T" transformers. This relationship is expressed as:

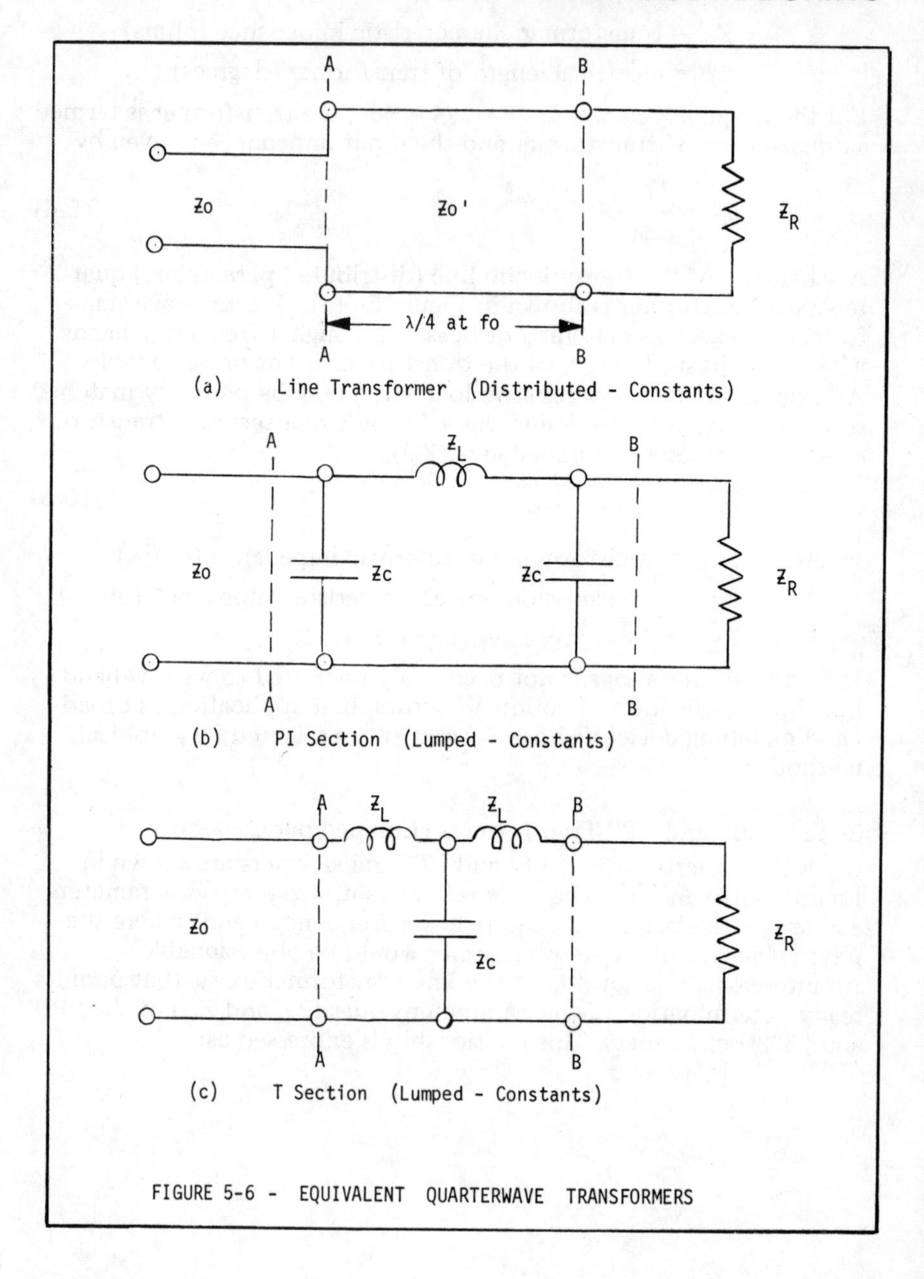

FIGURE 5-6 - EQUIVALENT QUARTERWAVE TRANSFORMERS

$$Z_L = +j\,Z_0' \tag{5-6}$$

$$Z_C = -j\,Z_0' \tag{5-7}$$

Where: Z_L = inductor impedance (ohms)

Z_C = capacitor impedance (ohms)

Z_0' = as determined by Equation (5-5) for the design frequency (fo)

To illustrate the compensation effects of the equivalent quarter-wave transformers that are shown in Figure 5-6, the following load impedances are selected to be compensated:

f (Relative)	Z Load	Z′ Load
$0.8\ f_0$	15 - j 10	0.3 - j 0.2
$1.0\ f_0$	15 + j 0	0.3 + j 0
$1.2\ f_0$	15 + j 10	0.3 + j 0.2

At the design frequency (f_0), the Z_0' of a line transformer for this case is:

$$Z_0' = \sqrt{50 \cdot 15} = 27.4 \text{ ohms}$$

and for the equivalent "PI" and "T" transformers the impedances are:

$$Z_L = +j\,27.4 \text{ ohms}$$

$$Z_C = -j\,27.4 \text{ ohms}$$

Figures 5-7, 5-8, and 5-9 show the compensation characteristics of the three equivalent quarter-wave transformers as applied to the same load impedances. Examination of the figures show that, at the design frequency (f_0), the compensation effects are the same and a perfect match, SWR = 1:1, is achieved. At frequencies other than the design frequency, the line transformer (Figure 5-7) has somewhat greater bandwidth potential as a result of the uniform transformation properties that are characteristic of distributed parameter elements.

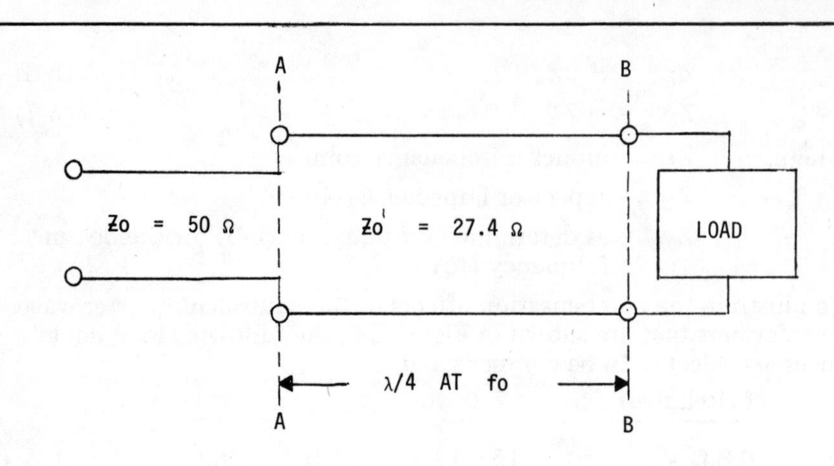

f	Ƶ LOAD	Ƶ LOAD (NORMALIZED TO Ƶo = 50)
0.8 fo	15 - j 10 Ω	0.3 - j 0.2
1.0 fo	15 + j 0 Ω	0.3 + j 0
1.2 fo	15 + j 10 Ω	0.3 + j 0.2

SMITH CHART PLOT OF IMPEDANCE AT REFERENCE PLANE A-A:

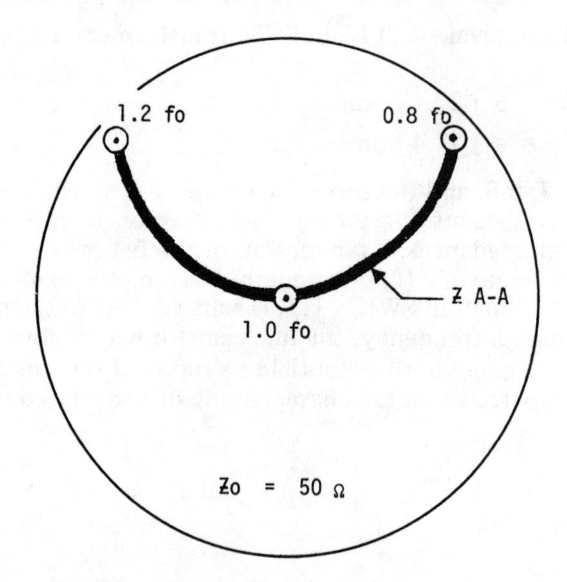

FIGURE 5-7 | IMPEDANCE MATCHING USING A QUARTERWAVE LINE TRANSFORMER

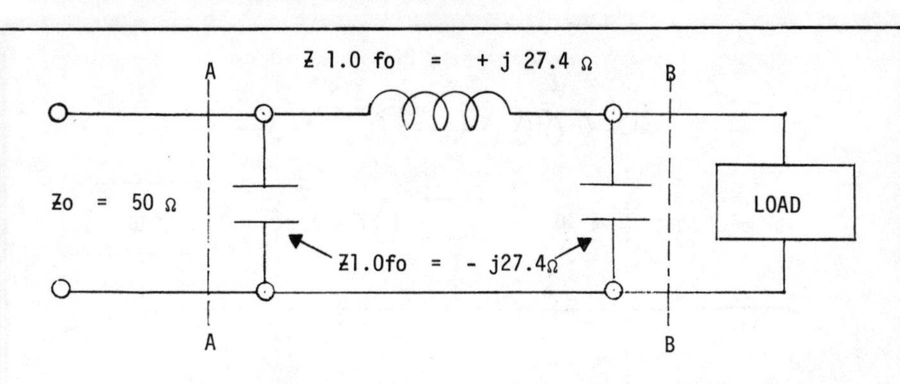

f	Z LOAD	Z LOAD (normalized to Zo = 50Ω)
0.8 fo	15 - j 10 Ω	0.3 - j 0.2
1.0 fo	15 + j 0 Ω	0.3 + j 0
1.2 fo	15 + j 10 Ω	0.3 + j 0.2

SMITH CHART PLOT OF IMPEDANCE AT REFERENCE PLANE A-A:

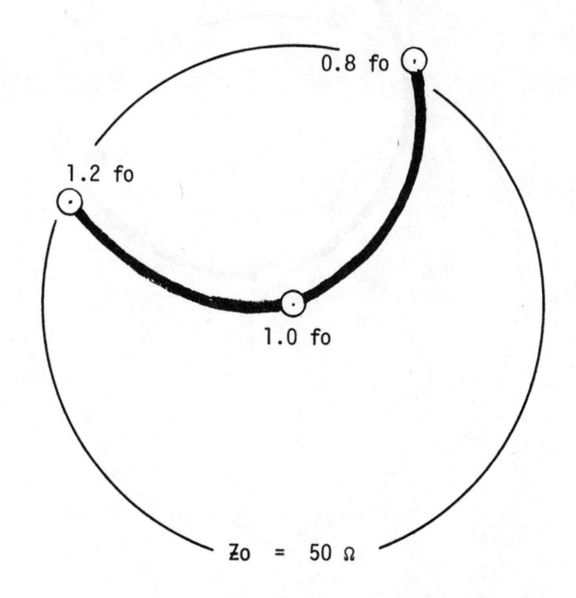

FIGURE 5-8 | IMPEDANCE MATCHING USING AN EQUIVALENT QUARTERWAVE PI-NETWORK

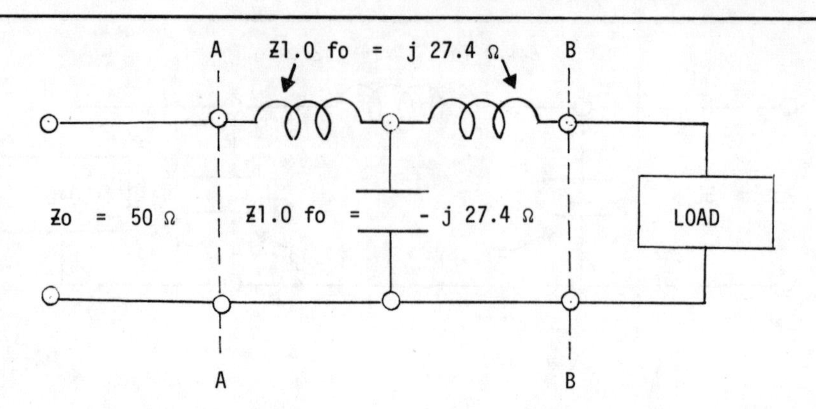

f	Z LOAD	Z LOAD (normalized to Zo = 50Ω)
0.8 fo	15 - j 10 Ω	0.3 - j 0.2
1.0 fo	15 + j 0 Ω	0.3 + j 0
1.2 fo	15 + j 10 Ω	0.3 + j 0.2

SMITH CHART PLOT OF IMPEDANCE AT REFERENCE PLANE A-A:

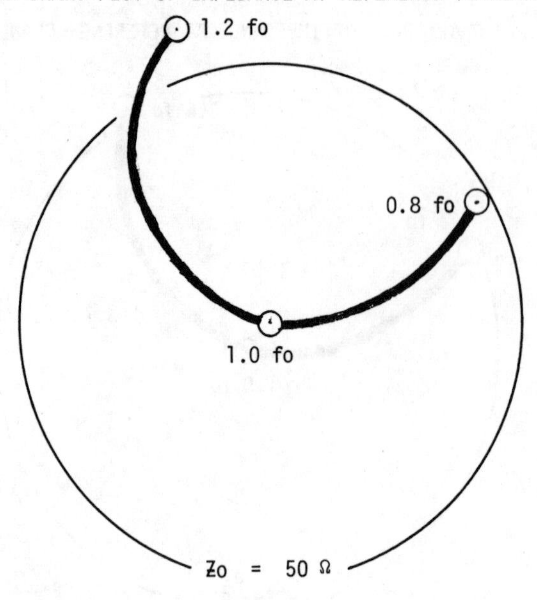

FIGURE 5-9 | IMPEDANCE MATCHING USING AN EQUIVALENT QUARTERWAVE T-NETWORK

5.5 LABORATORY TUNING DEVICES

5.5.1 Stub Tuners

In paragraph 3.3, open-circuited and short-circuited line segments were described as impedance matching devices. They behave as pure reactances whose input impedance values are readily determined by graphical construction on the Smith Chart. In the r.f. laboratory, line segments are more commonly termed "stub tuners" and are used as "in-line" matching devices both in the coaxial and waveguide forms. The majority of stub tuners are shorted stubs. Open-circuited stubs are rarely used as a true open circuit is difficult to achieve. Also an open line radiates energy which is not desirable in most laboratory test setups.

In Figure 5-10 are shown the schematics of both single and double stub tuners. Since the stubs are shunted across the transmission line, it is more convenient to use admittances to describe the compensation effects. A single stub must be located at the proper position in the transmission line in order to be effective. For example, consider the load $Y_L = G_L + j\,B_L$, located at Reference Plane A-A of Figure 5-10(a). If Y_L is rotated along the transmission line to a point on the Smith Chart where $Y_L = 1 + j\,B_L$, then a single stub inserted at this point will have the maximum compensation effectiveness. The shorting plunger of the stub may be adjusted to cancel out the susceptance ($\pm\, j\,B_L$) of Y_L to give $Y_L = 1 \pm j\,0$. Thus a perfect match is achieved and there will be no reflections on the transmission line. In practice it is usually inconvenient to locate a single stub at its proper position in the line. Double stub tuners, however, may be generally located at most any position in the transmission line and still cancel out most reflections. The distance between the two fixed stubs is usually $3/8\ \lambda$. To describe the compensation "action"

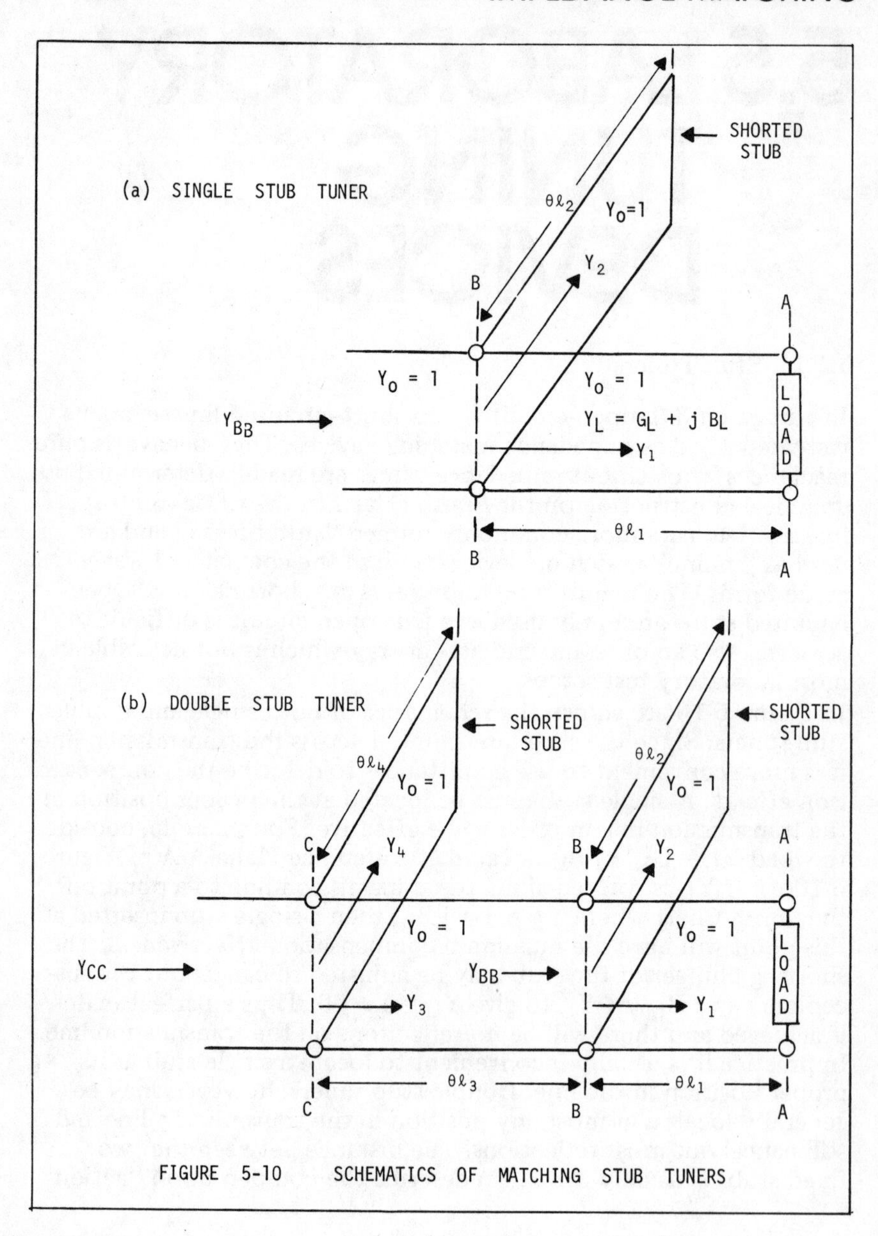

FIGURE 5-10 SCHEMATICS OF MATCHING STUB TUNERS

of a double stub tuner, consider the construction shown in Figure 5-11. In the figure, the admittance, Y_1, looking towards the load at Reference Plane B-B is 0.5 + j 1.5. The stubs are adjusted as follows:

1/The shorting plunger of the first stub is adusted to make $\theta \ell_2$ = .087 λ. This gives Y_2 = - j 1.63. Y_{BB} is then:

$$Y_{BB} = Y_1 + Y_2$$
$$= 0.5 + j\,1.5 + (-1.63)$$
$$= 0.5 - j\,0.13$$

In the above, Y_{BB} is positioned on the chart such that the distance between the stubs (3/8 λ) places the admittance, Y_3, on the G = 1.0 axis. At this point Y_3 = 1.0 - j 0.7.

2/The electrical length, $\theta \ell_4$, of the second stub is adjusted such that Y_4 = + j 0.7. The final input admittance, Y_{CC}, becomes:

$$Y_{CC} = Y_3 + Y_4$$
$$= 1.0 = j\,0.7 + j\,0.7$$
$$= 1.0 \pm j\,0 \text{ (a perfect match)}$$

The above example shows the theoretical aspects of matching with a double stub tuner. In the laboratory, this is not the case. The stubs are adjusted alternately, without resorting to calculations, until the SWR on the transmission line is reduced to unity.

5.5.2 Slide Screw Tuners

Slide-screw tuners are particularly useful for reducing reflections in waveguide transmission line test setups. This type of tuner consists of slotted section of waveguide on which is mounted a precision carriage containing an adjustable probe. The length of the slotted section is generally several wavelengths. The penetration of the probe into the guide is controlled by a micrometer drive whereas the probe position, along the guide, is varied by a thumb-operated wheel. Thus the depth and position of the probe may be varied to setup a reflection to cancel an existing reflection that is existing on the line.

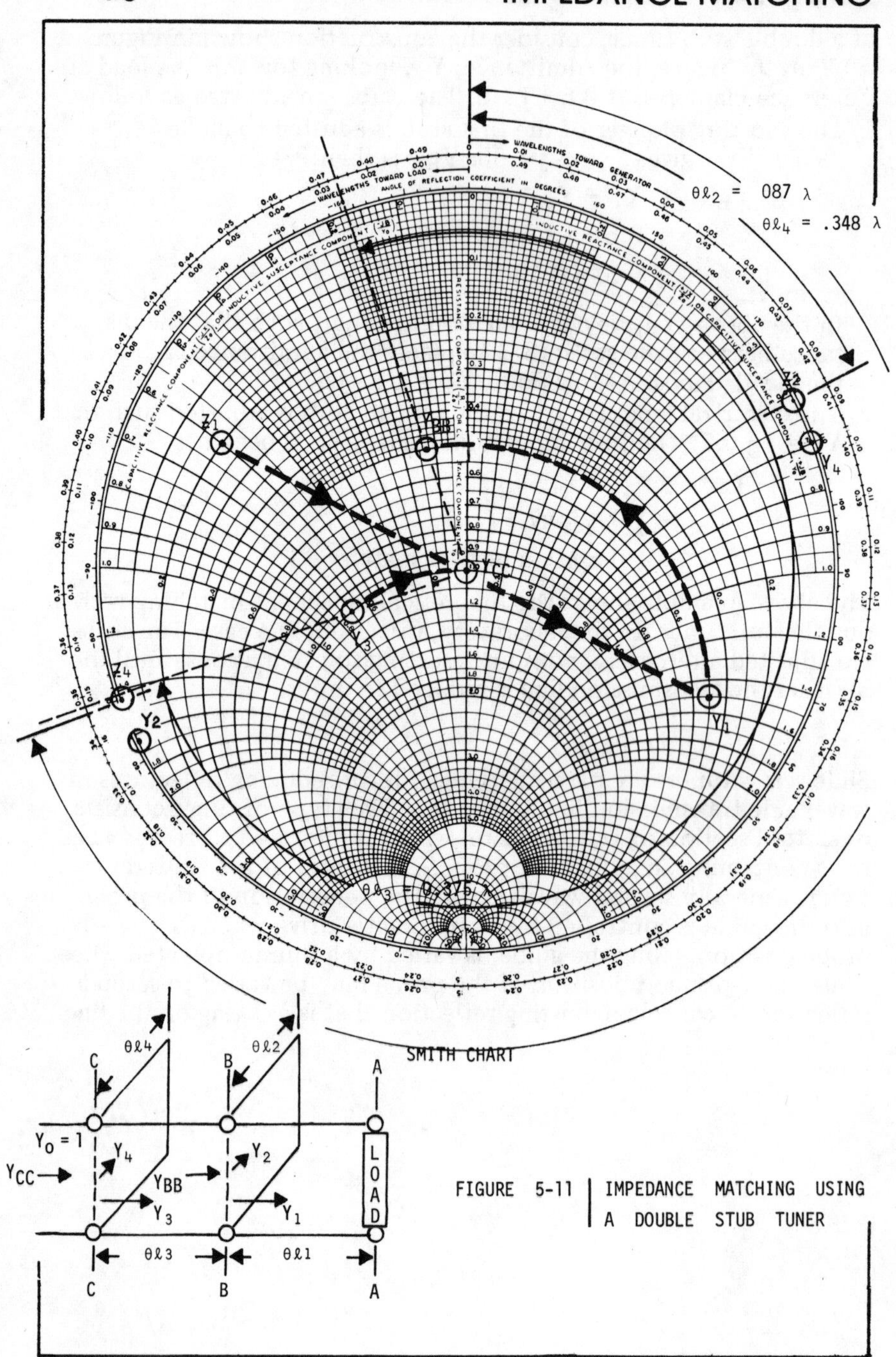

FIGURE 5-11 | IMPEDANCE MATCHING USING A DOUBLE STUB TUNER

5.5.3 Hatch Tuner[3]

Another interesting transmission line tuning device is the Hatch Tuner whose coaxial representation is shown in Figure 5-12. The transmission line may be bent in a circle, Reference Planes A-A and D-D near coinciding, to provide a device consisting of series and shunt sections of transmission lines whose lengths are adjustable by two moving shorting sections. The Hatch Tuner was initially developed to be used in scale model aircraft, such as that shown in Figure 5-13, to provide impedance compensation for model antennas operating in the 300 MHz to 1000 MHz frequency bands.

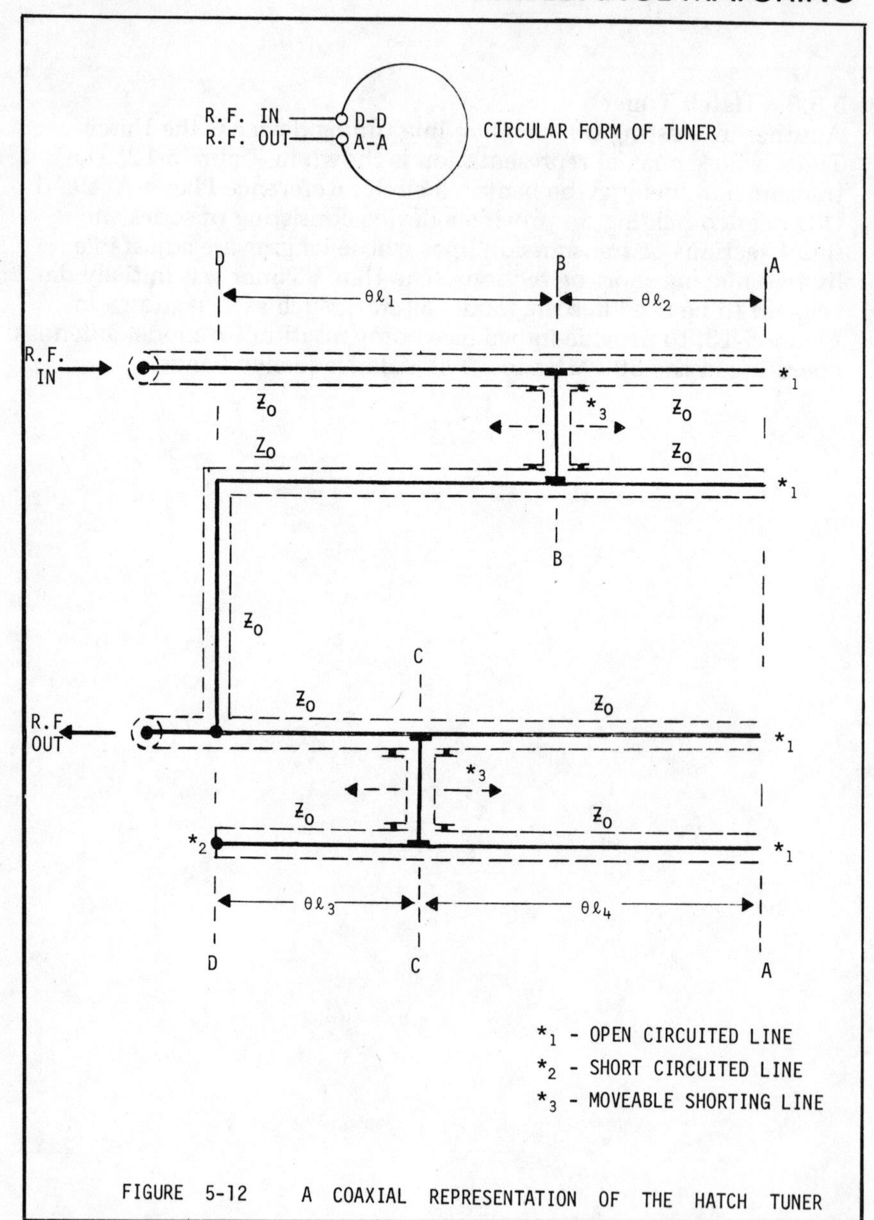

FIGURE 5-12 A COAXIAL REPRESENTATION OF THE HATCH TUNER

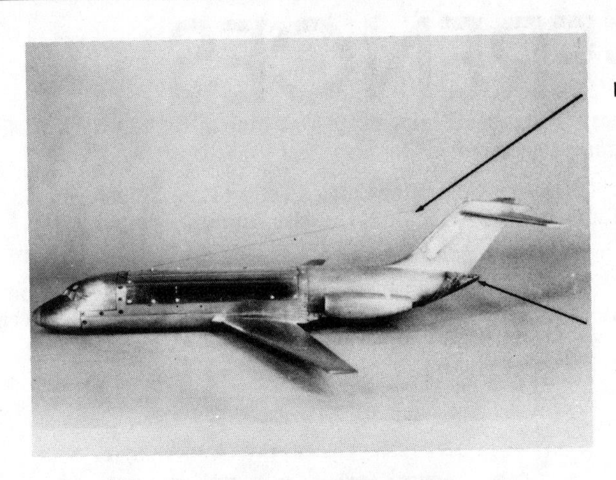

A 1/30th Scale Model Aircraft
for Radiation Pattern Studies

FIGURE 5-13 | HATCH TUNER — A VARIABLE MATCHING DEVICE USED IN MODEL PATTERN STUDIES

REFERENCES

1. R.L. Thomas, "Broadband Impedance Matching in High-Q Networks," *EDN Magazine*, December 1973.

2. R.L. Thomas, "Design Charts for Single Section Line Transformers," Technical Report No. DAC 66662, Douglas Aircraft Company, Inc., Long Beach, California. 1967.

3. R.M. Hatch, "Improvements in Instrumentation for the Investigation of Aircraft Antenna Radiation Patterns by means of Scale Models," Technical Report No. 26, Stanford Research Institute, dated August 1952.

BIBLIOGRAPHY

1. G.J. Wheeler, *Introduction to Microwaves*, Prentice-Hall Book Co., Inc., New Jersey, 1963.

2. J.A. Nelson and G. Stavis, "Impedance Matching, Transformers and Baluns," *Very High Frequency Techniques*, Chapter 3, Radio Research Laboratory, McGraw-Hill Book Company, New York and London. 1947.

3. H.F. Mathis, "L-Network Design," *Electronics Magazine*, pp 186-188, dated 1 February 1957.

6 BROADBAND MATCHING—LINE TRANSFORMER DESIGN CHARTS

6.1 INTRODUCTION

The term "broadband matching", as defined in Section V, implies impedance compensation where bandwidths are greater than 50%. This type of matching is more technically complex when compared to single frequency or narrow band matching. The previous section has treated simple networks such as (a) single element, (b) two-element, and (c) quarterwave transformers where the impedance bandwidths of concern were narrow to moderate. In these cases, very little technical effort need be expended to achieve a satisfactory match. Broadband matching, however, may require considerable technical effort and the degree of success is directly related to the ingenuity of the designer in his selection of appropriate networks.

Two sections are here devoted to the discussion of broadband matching techniques. The present section describes the use of design charts, as graphical aids, in the derivation of design parameters for broadband line transformers. Such aids permit a rapid accessment of the compensation effects of a given transformer design. Section VII will discuss the use of four-element optimum networks for maximum bandwidth. Such networks are particularly applicable for compensating high "Q" structures.

6.2 LINE TRANSFORMER ELECTRICAL PARAMETERS

A transformer that is inserted in series with a load and the associated r.f. transmission line is termed a "line transformer." Principal parameters of such a transformer is its electrical length ($\theta \ell$) and characteristic impedance (Z_0'). In a given design Z_0' remains constant whereas $\theta \ell$ varies with frequency.

Each specific transformer has an associated family of transformation circles or θ circles. The significance of a particular θ circle is that the circle is transformed to the definition circle (SWR) when the transformer electrical length equals θ. θ circles, plotted upon design charts, provide a method for performing a graphical analysis of the compensation characteristics of a given transformer design. A description of such design charts and their application follows.

6.3 DESIGN CHARTS

6.3.1 General

Tabulated data to permit the formulation of design charts for line transformers for various SWR values is reported in Reference 1. Design charts have been compiled from this data for the case where the definition circle (SWR) equals 2:1 and are included in Appendix A. These charts permit a rapid analysis of line transformer characteristics.

Figure 6-1 shows a typical design chart. The chart coordinate system (resistance-reactance axis) and transformer characteristic impedance (Z_0') have been normalized to an associated transmission line characteristic impedance (Z_0) of 1.0. The charts are thus universal in application and not limited to the specific case of Z_0 = 50 ohms. In Appendix A, charts are provided with transformer characteristic impedance values in 0.1 Z_0 increments from 0.4 Z_0 to 3.0 Z_0.

6.3.2 Chart Description

- *Title Block* — The SWR ratio and normalized value of Z_0 and Z_0' are given here. Should the characteristic impedance (Z_0) of the associated line be 50 ohms, this particular chart would reflect the design parameters of a 35 ohm (0.7 Z_0) transformer (Figure 6-1).
- *Coordinate System* — The resistance-reactance axis are normalized to Z_0 = 1.0. Since the impedances plotted on the Smith Chart are in normalized form, they may be directly transferred to the design chart.
- *Definition Circle (SWR)* — The definition circle, or SWR circle, is established by performance specifications and all impedance values contained within the boundaries of this circle are considered to be matched. The SWR 2:1 circle crosses the normalized resistance axis at 0.5 and 2.0.

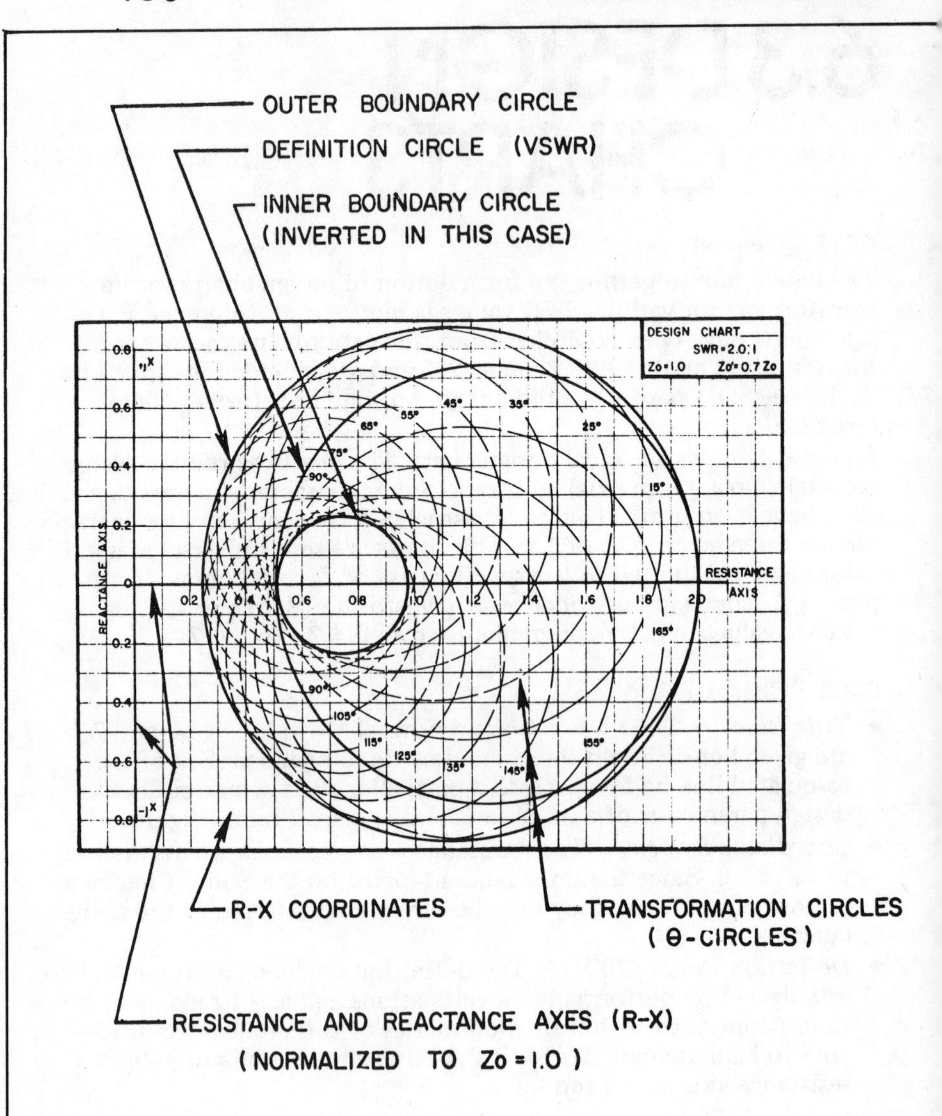

FIGURE 6-1 TYPICAL DESIGN CHART FOR SINGLE SECTION LINE TRANSFORMER

- *Outer Boundary Circle* — This circle encloses the SWR circle and all points on the impedance curve that is to be matched. The circle is tangent to the SWR circle, on the resistance axis at 2.0 when $Z_0' <$ than Z_0 and 0.5 when $Z_0' > Z_0$.
- *Inner Boundary Circle* — The inner boundary circle is tangent to both the $\theta = 90°$ circle and the SWR circle. This circle is said to be inverted when it is contained within the SWR circle.
- *Transformation Circles (θ Circles)* — Each transformer has an associated family of transformation circles. All θ circles are different in size and progress in a counterclockwise manner on the charts. All θ circles are tangent to (a) the outer boundary circle and (b) the inner boundary circle. It is the characteristics of the θ circles that provides the method of performing a graphical analysis of line transformer matching susceptibility.

6.4 ANALYSIS FACTORS IN DESIGN SELECTION

Simplicity of design chart usage is based upon an understanding of the following fundamental analysis factors:

- A line transformer has different electrical lengths ($\theta \ell$) at each frequency considered. Each transformer has a specific family of θ circles. The transformer will transform to the SWR circle one θ circle at one frequency and a different θ circle at another frequency.

- When an impedance point falls within a θ circle, the point is transformed to within the SWR circle at the frequency where the electrical length of the transformer equals to θ. The degree of compensation afforded by a θ circle is illustrated in Figure 6-2. Should the impedance point lie near the boundary of the θ circle, it will be transformed to a position near the boundary of the SWR circle. The impedance point will be transformed to near the center of the SWR circle, if it is initially near the center of the θ circle.
- If successive points on the impedance curve fall within the respective θ circles for those frequencies, the entire impedance curve may be transformed to within the SWR circle by that particular transformer. This implies that increasing frequency impedance points should "track" increasing θ circles for effective matching.
- The portion of the impedance curve that lies in the region between the inner and outer boundary circles is susceptible to matching. Points outside the outer boundary circle or within the inner boundary circle (except when this circle is inverted) are not susceptible to matching by that particular transformer.
- Impedance points within an inverted inner boundary circle remain in that circle (and hence, within the SWR circle) regardless of the electrical length of the transformer. Such points are, therefore, of no concern in the analysis.
- In the selection of a trial electrical length ($\theta \ell$) for the transformer for matching analysis, consideration should be given to the critical points of the impedance curve. Critical points are those that (a) lie near the outer boundary circle or (b) on a portion of the impedance curve that moves in a clockwise manner (direction of decreasing θ circles) with increasing frequency.

There are two basic electrical characteristics that define a particular line transformer; the characteristic impedance (Z_0') and electrical length ($\theta \ell$). Z_0' is derived by simple construction methods using the Smith Chart, whereas $\theta \ell$ is selected by trial by judicious consideration of the above analysis factors. Both processes are described by the selected example of chart usage that follows.

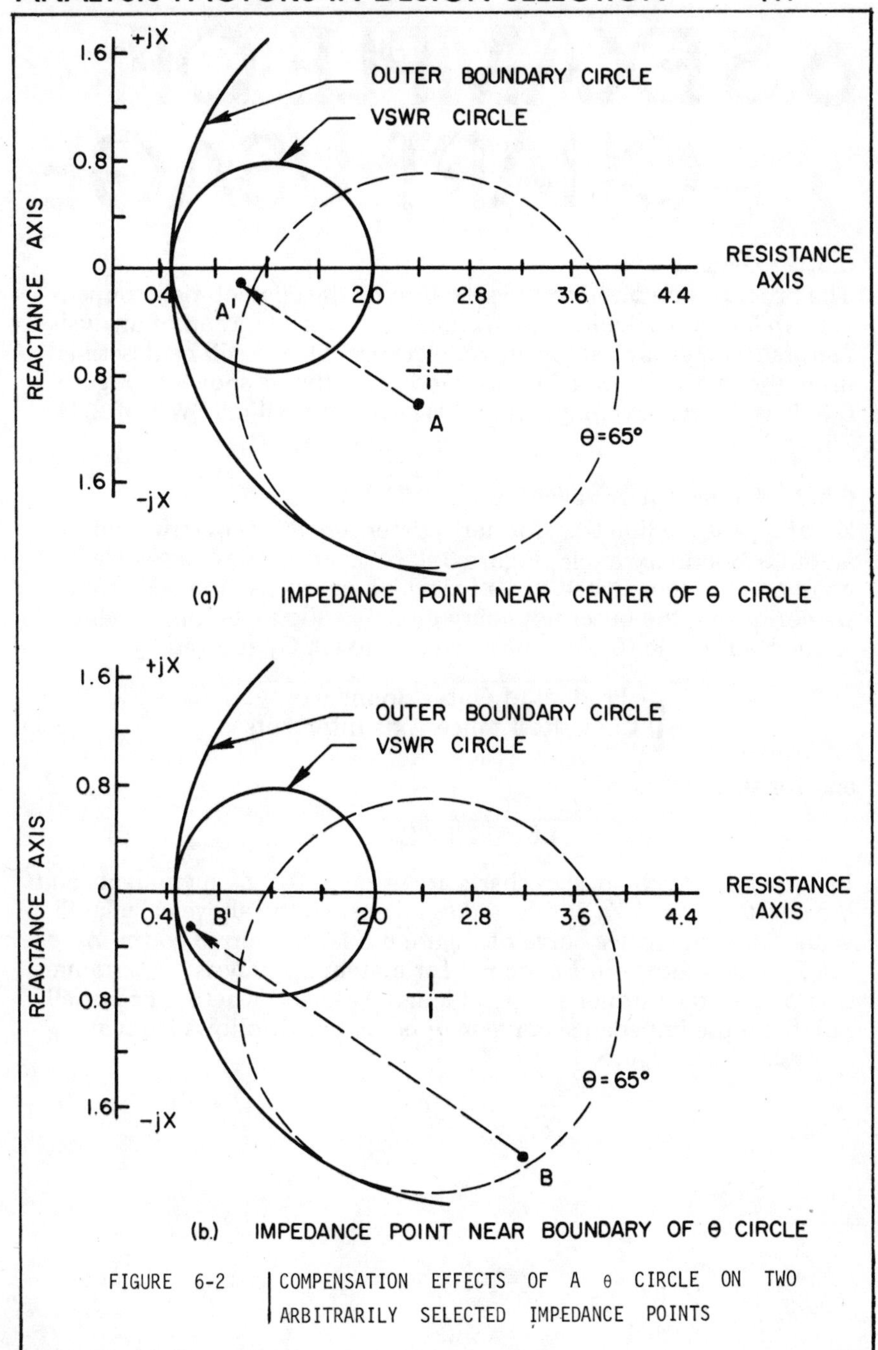

FIGURE 6-2 COMPENSATION EFFECTS OF A θ CIRCLE ON TWO ARBITRARILY SELECTED IMPEDANCE POINTS

6.5 EXAMPLE OF CHART USAGE

6.5.1 General

The processes involved in the selection of the characteristic impedance (hence a particular design chart) and consideration of analysis factors to determine an optimum electrical length will be described using the uncompensated impedance curve that is shown in Figure 6-3. It is desired to compensate this curve to within a SWR of 2:1.

6.5.2 Design Chart Selection

Z_0' of a possible line transformer is determined by construction of an outer boundary circle, Figure 6-3, by inspection to enclose all points on the impedance curve for which compensation is desired. By definition, the outer boundary circle, in Figure 6-3, is tangent to the SWR circle (0.5 on the resistance axis). Z_0' is given by:

$$Z_0' = \sqrt{\begin{matrix}\text{Product of Outer Boundary}\\ \text{Circle Resistance Axis Intercepts}\end{matrix}} \tag{6-1}$$

and for this case is:

$$Z_0' = \sqrt{0.5 \times 4.6} = 1.52\ Z_0$$

In Appendix A, the design charts are given in 0.1 Z_0 increments and the chart $Z_0' = 1.5\ Z_0$ is selected as nearest to the above calculated value. The impedance curve of Figure 6-3 is transferred to the $Z_0' = 1.5\ Z_0$ chart shown in Figure 6-4 for matching analysis. Determination of the transformer electrical length ($\theta \ell$) and whether or not all points on the impedance curve may be matched follows from an analysis of this figure.

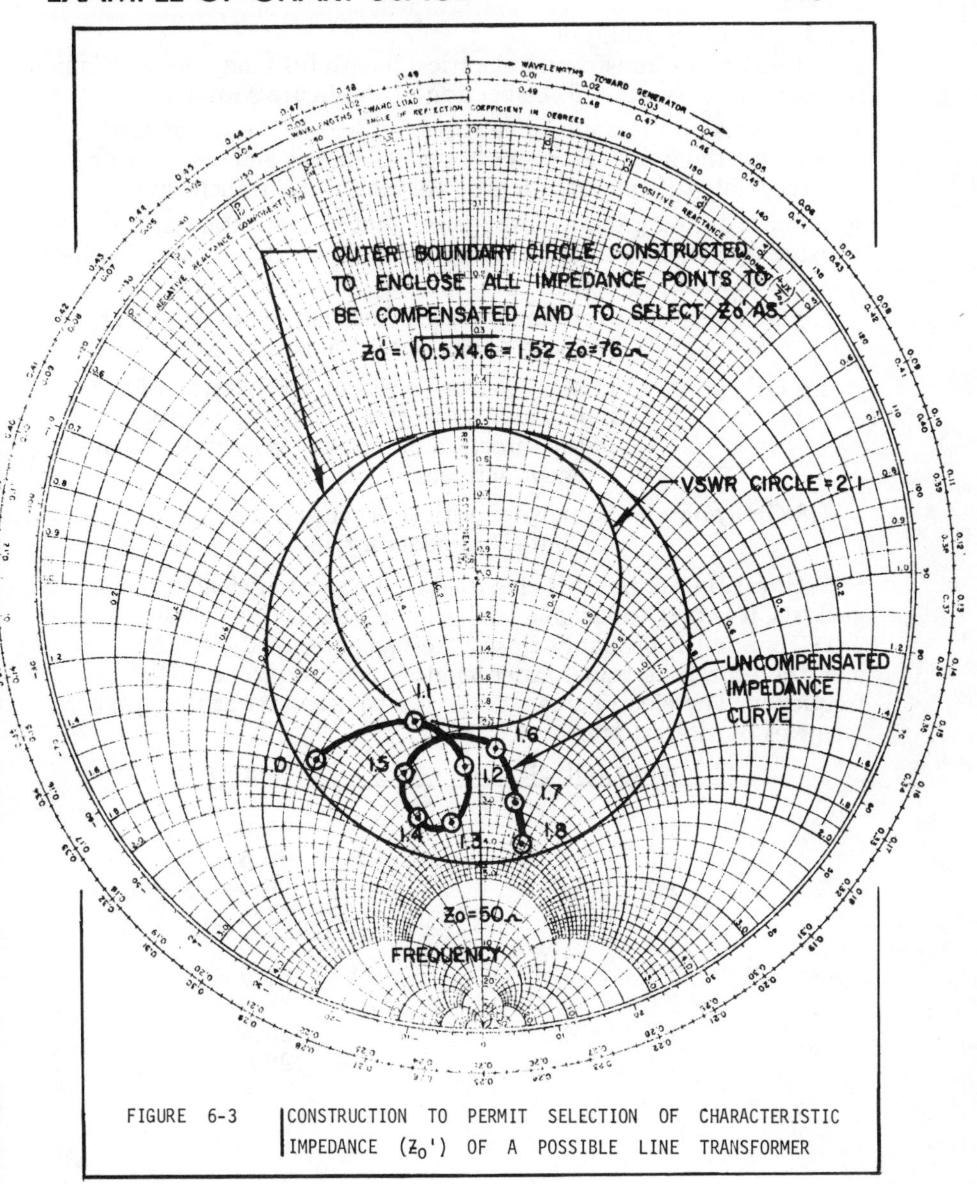

FIGURE 6-3 CONSTRUCTION TO PERMIT SELECTION OF CHARACTERISTIC IMPEDANCE (Z_0') OF A POSSIBLE LINE TRANSFORMER

6.5.3 Matching Analysis

Selection of the transformer electrical length ($\theta \ell$) may be determined by analysis of Figure 6-3. Inspection of the figure shows:

- All points on the impedance curve fall within the region that is between the inner and outer boundary circles and are, therefore, susceptible to compensation by this particular transformer.
- There are three points on the curve that fall into the critical points category. These points and reasons for their critical classification are as follows:

  1.0 GHz point — Extremely critical because of the very near position to the outer boundary circle.

  1.4 GHz point — Critical because it lies on the portion of the impedance curve that is starting to progress in the direction of decreasing θ circles as frequency increases.

  1.8 GHz point — Somewhat critical because of its position near the outer boundary circle.

Since the 1.0 GHz point is classified as the most critical in the above analysis, it is best to assign a trial value for $\theta \ell$ at this frequency. The 1.0 GHz point is probably closest to the center of the $\theta = 55°$ circle and, therefore, the transformer length is selected to be 55° at this frequency. The electrical lengths of the transformer at the other frequencies of interest are:

Frequency (GHz)	$\theta \ell$ (degrees)
1.0	55.0
1.1	60.5
1.2	66.0
1.3	71.5
1.4	77.0
1.5	82.5
1.6	88.0
1.7	93.5
1.8	99.0

By definition, an impedance point lying within its respective θ circle for that frequency will be transformed to within the SWR circle. Examination of Figure 6-4 shows that all impedance points fall within their respective circles. The entire impedance curve may be compensated to within the SWR circle by a transformer whose $Z_0' = 1.5\ Z_0$ and $\theta \ell = 55°$ at 1.0 GHz. Assuming that the associated transmission line $Z_0 = 50$ ohms, the Z_0' of the transformer will be 75 ohms. The compensated impedance curve is shown in Figure 6-5. This curve is derived from methods given in Section III and verifies the above design selection achieved by graphical analysis.

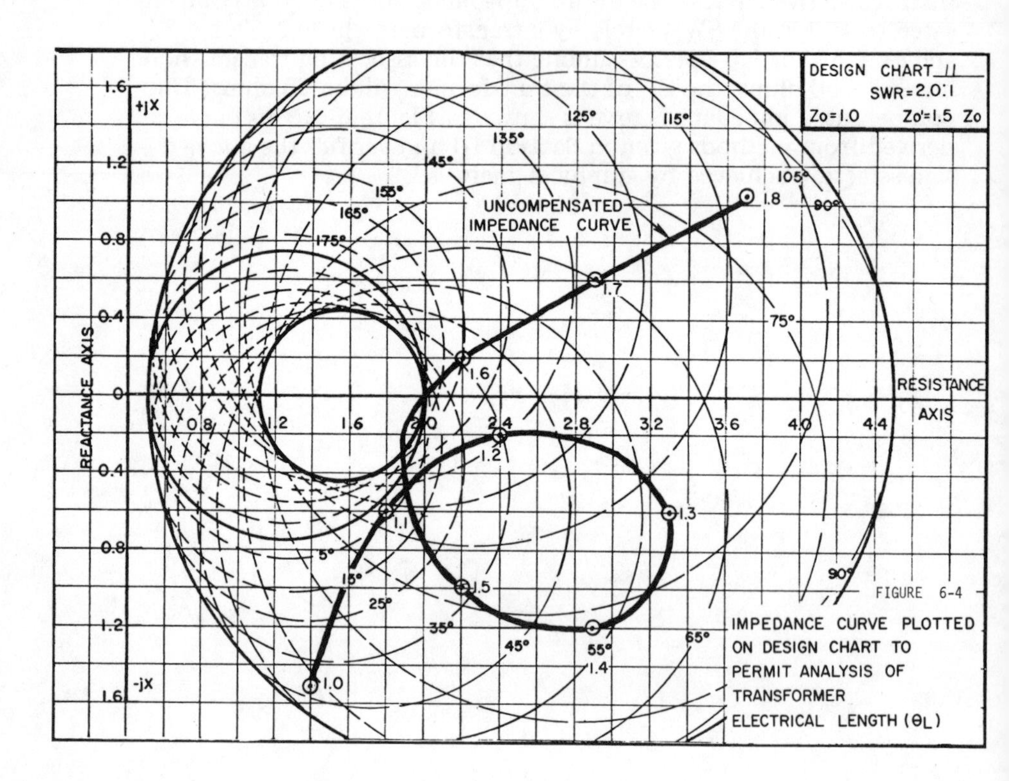

FIGURE 6-4

IMPEDANCE CURVE PLOTTED ON DESIGN CHART TO PERMIT ANALYSIS OF TRANSFORMER ELECTRICAL LENGTH (θ_L)

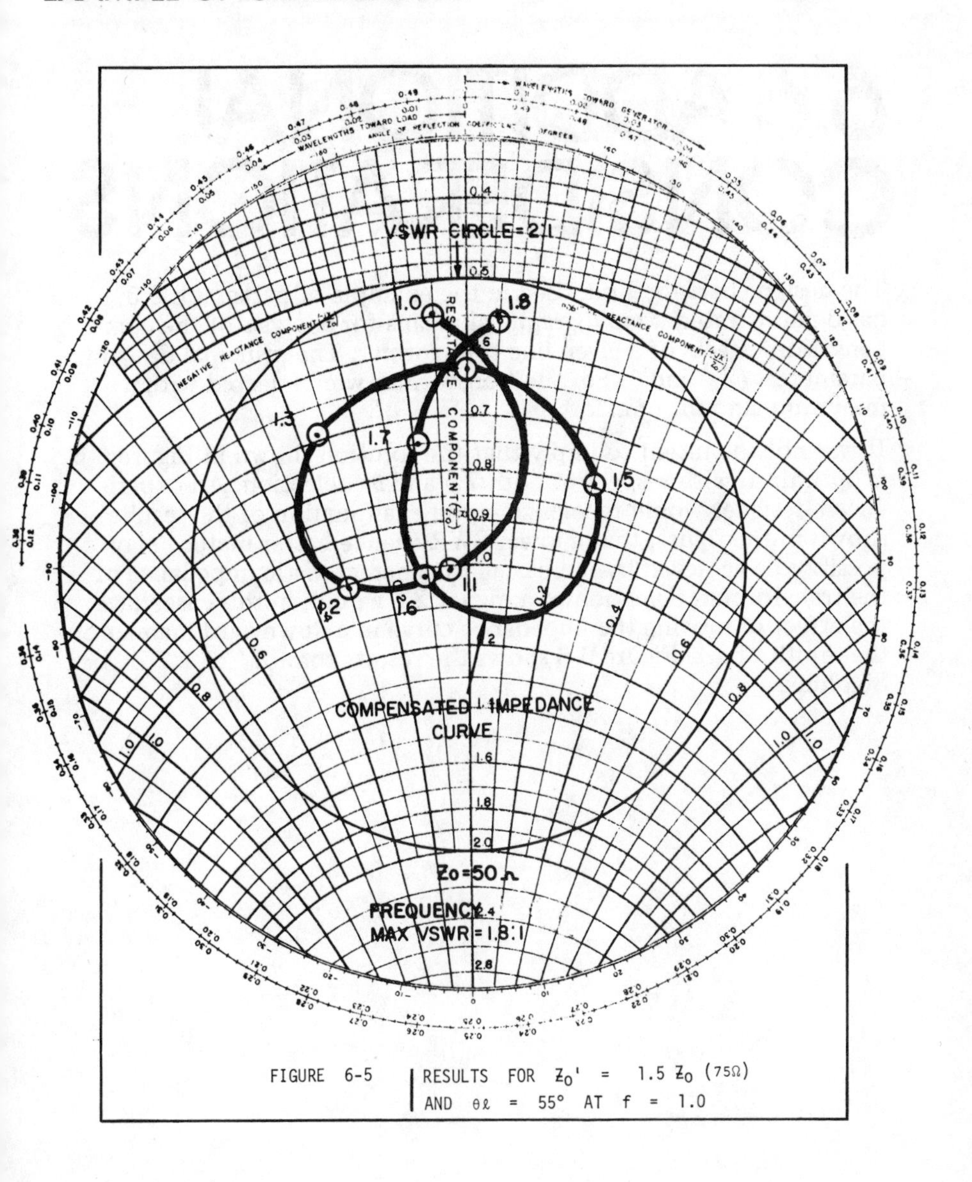

FIGURE 6-5 RESULTS FOR $Z_0' = 1.5\ Z_0$ (75Ω) AND $\theta\ell = 55°$ AT $f = 1.0$

6.6 ADDITIONAL CONSIDERATIONS

The design chart method, used in the example of paragraph 6.5, has demonstrated a rapid graphical means for analysis of the compensation effects of a given line transformer. The principal design parameters (Z_0' and $\theta \ell$) of the transformer were selected with a minimum amount of effort.

It was not the intent to imply that the solution shown in Figure 6-5 is optimized with respect to both Z_0' and $\theta \ell$ selection. An experienced designer would recognize, upon examination of the results shown in the figure, that a lower SWR (hence better match) may possibly be achieved by increasing the characteristic impedance of the transformer. Selection of a higher Z_0' would, in this case, have the effect of moving the impedance curve in a downward manner on the Smith Chart. Figure 6-6 shows the results for a Z_0' of 1.7 Z_0 or 85 ohms.

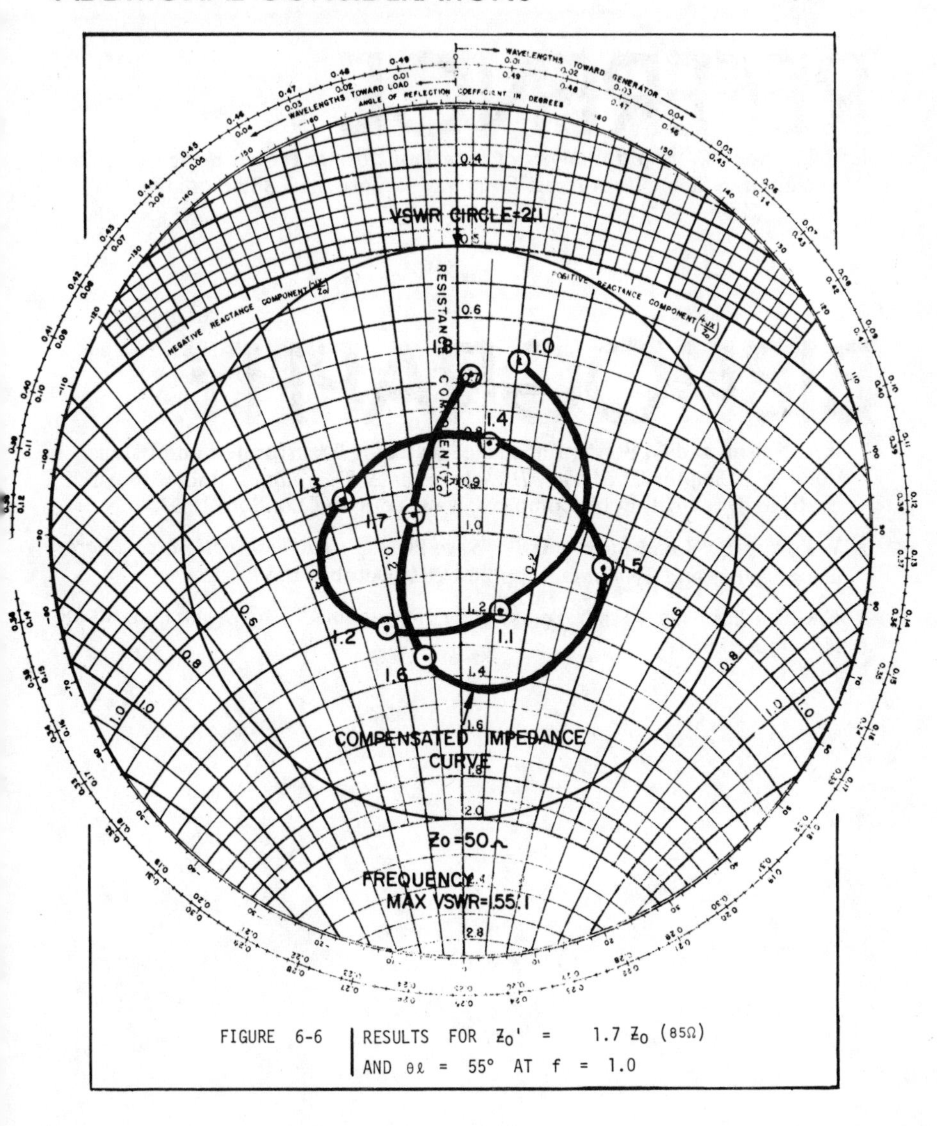

FIGURE 6-6 RESULTS FOR $Z_0' = 1.7\ Z_0$ (85Ω) AND $\theta\ell = 55°$ AT $f = 1.0$

REFERENCE

1. R.L. Thomas, "Rapid Analysis of Line Transformer Design Using Computer Techniques," Douglas Aircraft Company Report DAC 33276, dated 11 August 1966.

BIBLIOGRAPHY

1. J.A. Nelson and G. Stavis, Impedance Matching Transformers and Baluns, Volume 1 Chapter III of *Very High Frequency Techniques*, Radio Research Staff, McGraw-Hill Book Company, New York and London, 1947.

2. W.P. Ayres, "Broad-Band Quarter-Wave Plates," *IRE Transactions on Microwave Theory and Techniques*, pp 258-261, October 1957.

3. F.E. Gardiol, "Chebyshev Transformer Nomograms," *Microwaves*, pp 56-58, November 1967.

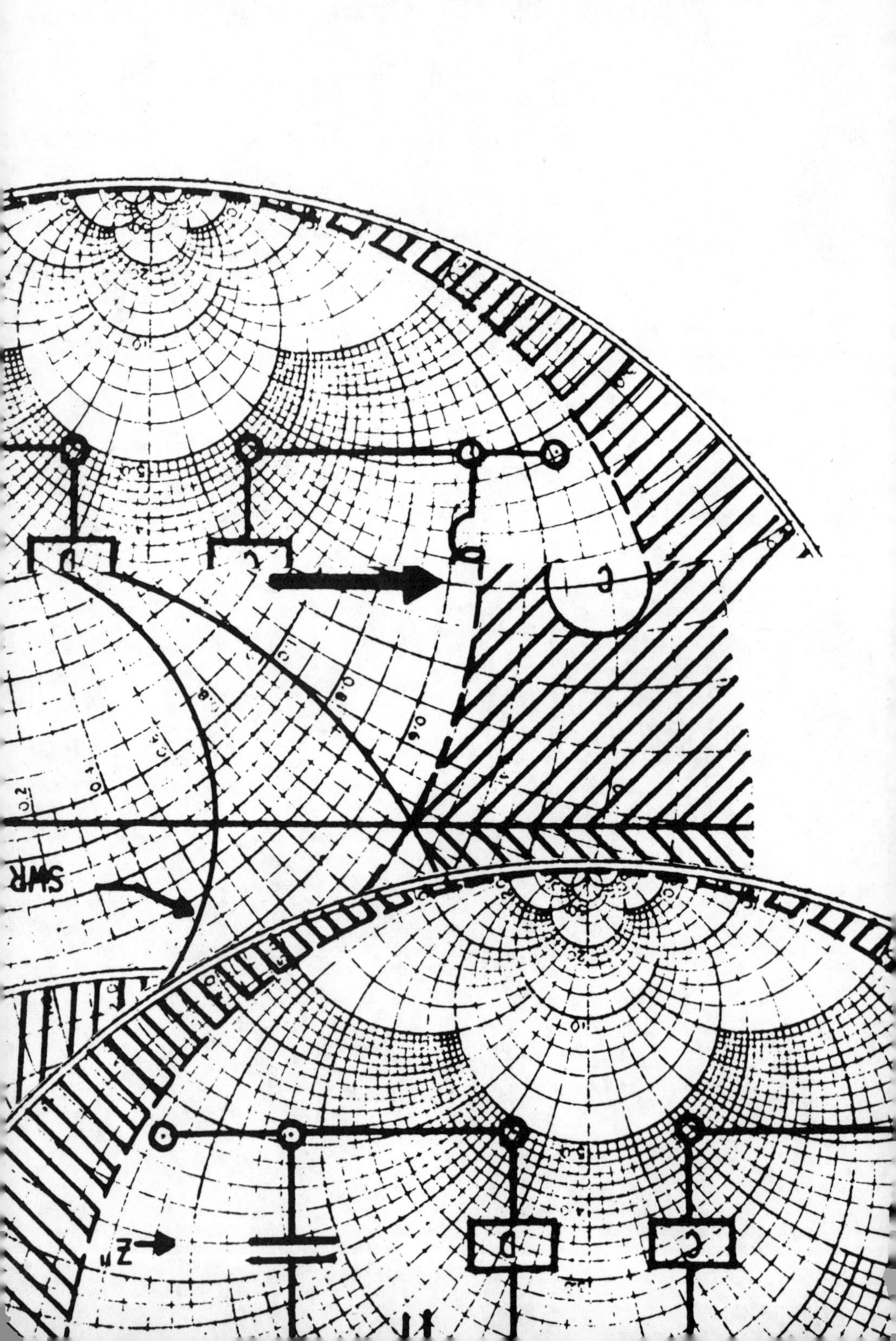

7 BROADBAND MATCHING—FOUR ELEMENT NETWORKS

7.1 INTRODUCTION

Section VI has discussed the application of the line transformer as a broadband device for impedance matching. Such transformers find particular application in the higher frequency bands where their physical length is relatively small and the impedance properties of the component to be compensated reflect moderate or better "Q" characteristics. In the lower frequency bands, however, it is not p tical to use the "distributed parameter" line transformer as a result of its excessive physical length and limited effectiveness in matching high "Q" structures.

This section will describe the application of "lumped parameter" four-element networks that are particularly applicable for matching high "Q" structures in the lower frequency bands. A theory of broadband matching is presented that permits the networks to be selected that will provide maximum impedance bandwidth. This is of importance since the more simple and least number of networks used in a given application the more efficient and less complex will be the resulting design.

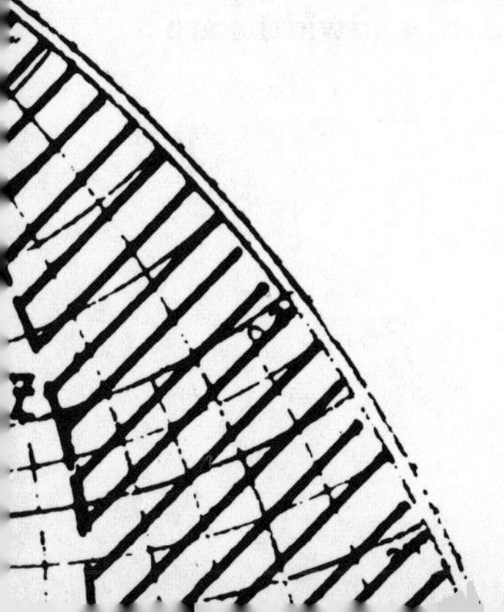

7.2 THEORY OF BROADBAND MATCHING

The theory of broadband impedance matching to be presented permits network selection to achieve maximum impedance bandwidth and the minimum number of fixed-tuned networks in a given design. A maximum of four elements only are considered in a particular fixed-tuned network selection as past theory[2] has suggested that very little bandwidth improvement is obtained for additional elements.

A statement of the theory of broadband matching is as follows:

- The first element of the four-element fixed-tuned network should be incorporated (a) as near as practical to the device terminals and (b) selected to reduce the input SWR of the initial impedance curve.
  Those familiar with Smith Chart operation are well aware that impedance points "spread out" as the electrical length from the device terminals is increased. Incorporation of networks far removed from the device terminals makes matching more difficult and results in reduced bandwidth potential.
  Proper network configuration selection (series or shunt) of the first element to reduce the initial SWR (hence a lower "Q") provides a better opportunity of obtaining broader bandwidths when the remaining network elements are added. Figure 7-1 is a graphical design aid that permits the proper network selection to match high "Q" impedances that may be located in various areas on the Smith Chart.
- In broadband applications, networks should never be selected to provide a perfect match for any one portion of a given impedance curve. Past experience has shown that broader bandwidths are

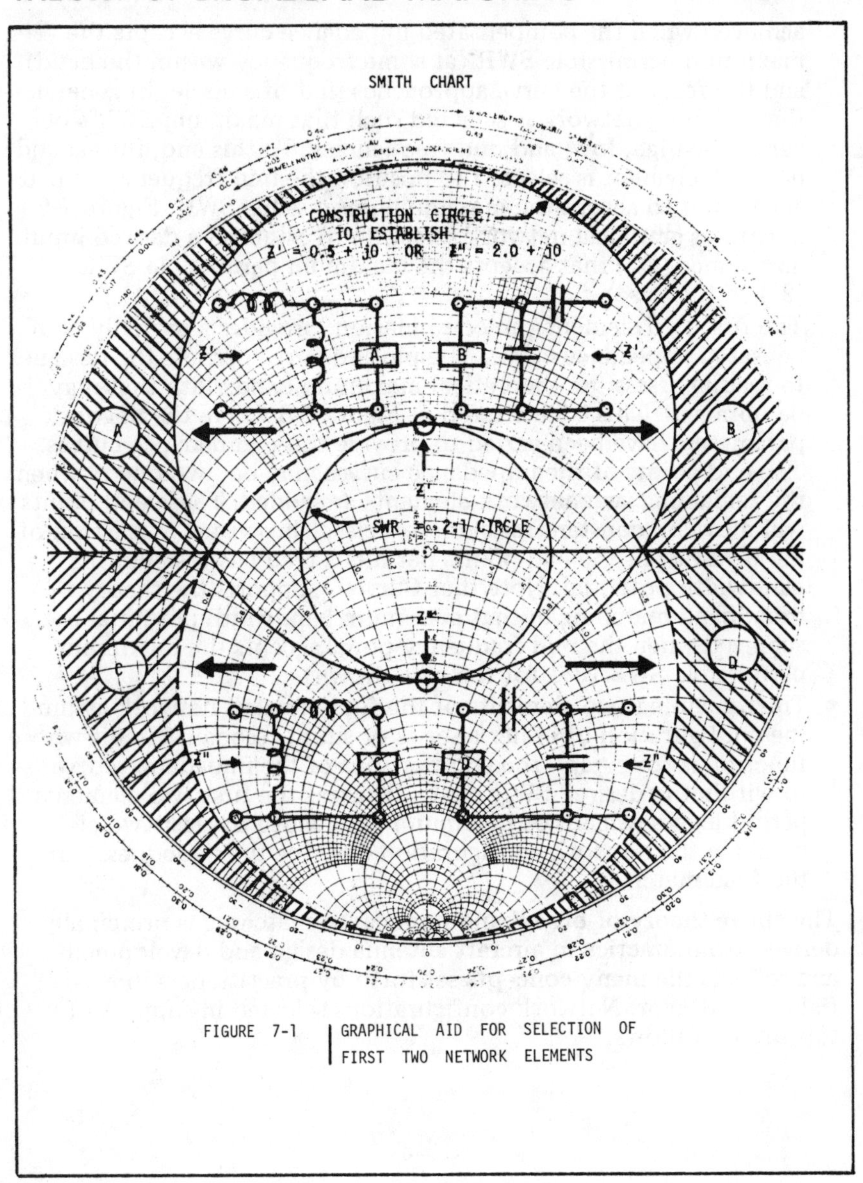

FIGURE 7-1 | GRAPHICAL AID FOR SELECTION OF FIRST TWO NETWORK ELEMENTS

achieved when the compensated impedance curve accepts the maximum permissible SWR (at some frequency within the band) and the form of the curve approaches that of a circle. In general, the matching network is selected such that maximum SWR's occur at the high, low, and mid frequencies. To this end, the second network element is selected to position the mid frequency impedance point to accept the maximum permissible SWR. Figure 7-2 illustrates graphical network selection to establish a desired input impedance (Z_c') that accepts the maximum permissible SWR (2:1 in this case).

The first two elements, in combination, provides a partially compensated impedance curve. The plotted curve permits the designer to readily assess the theoretical maximum bandwidth that may be obtained. This is illustrated in Figure 7-3 where the network parameters have been varied to provide three impedance curves. The end points of the impedance curves may be "wrapped" when the remaining two network elements are added. These end points will move on constant resistance circles. Thus visual inspection of the behavior of the end points permits a ready assessment of bandwidth potential. Note that Curve C supports the theory of broadband matching in that maximum bandwidth potential is achieved when the mid frequency point accepts the maximum permissible SWR (3:1 for the case shown).

- The remaining two elements of the four-element network assume the configuration of either a series or shunt resonant circuit whose function is to "wrap" the end portions of the impedance curve to within the desired SWR circle. Whereas the first two elements permit an assessment of maximum bandwidth the last two elements in the network configuration describe the "goodness" of the final match.

The above theory of broadband impedance matching is principally derived from practice in aircraft antenna design and development and reflects the many concepts set forth by practitioners in this field of endeavor. Network configurations selected in support of this theory follows.

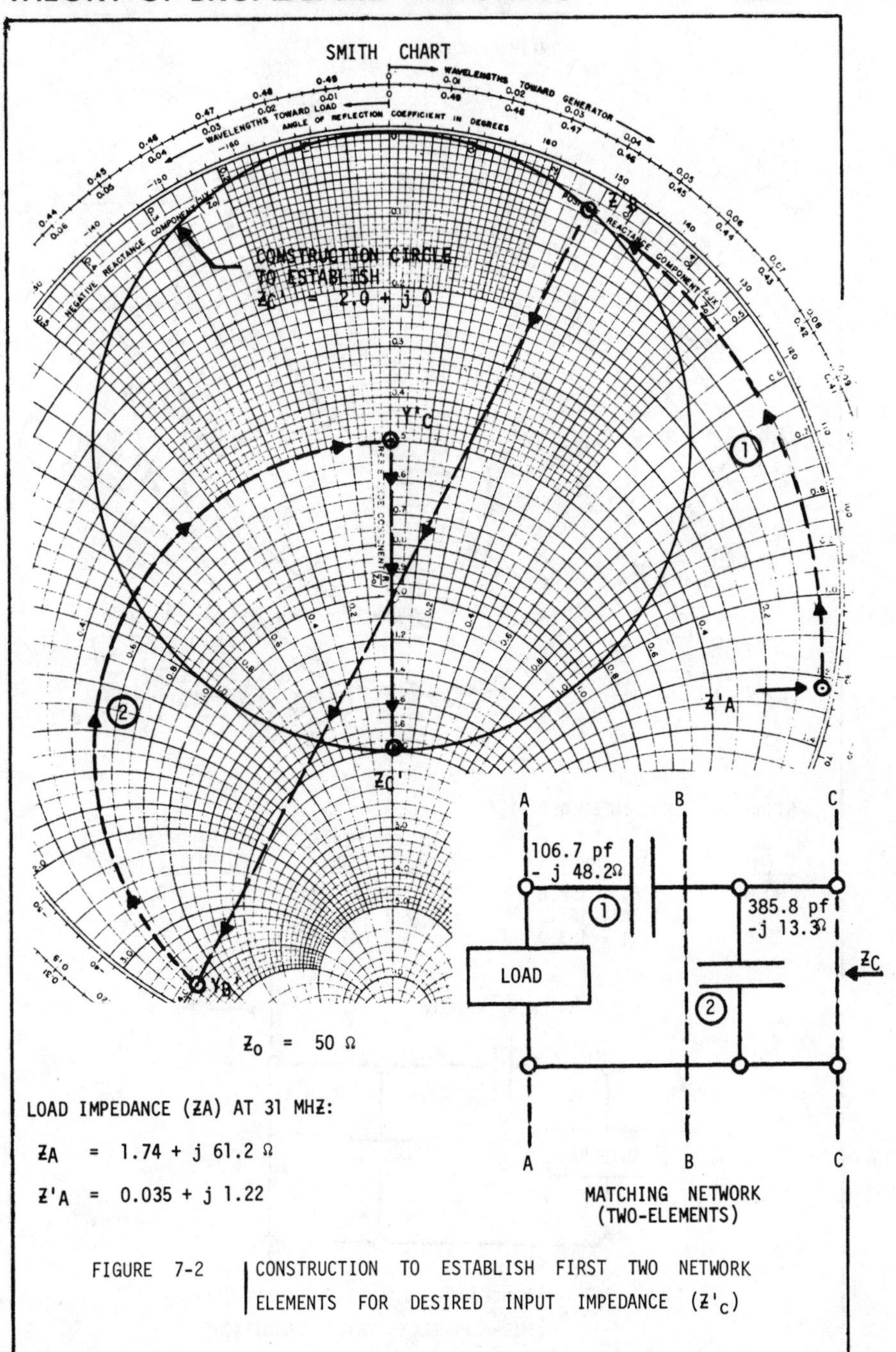

$Z_0 = 50\ \Omega$

LOAD IMPEDANCE (ZA) AT 31 MHZ:

$Z_A = 1.74 + j\,61.2\ \Omega$

$Z'_A = 0.035 + j\,1.22$

FIGURE 7-2 | CONSTRUCTION TO ESTABLISH FIRST TWO NETWORK ELEMENTS FOR DESIRED INPUT IMPEDANCE (Z'_C)

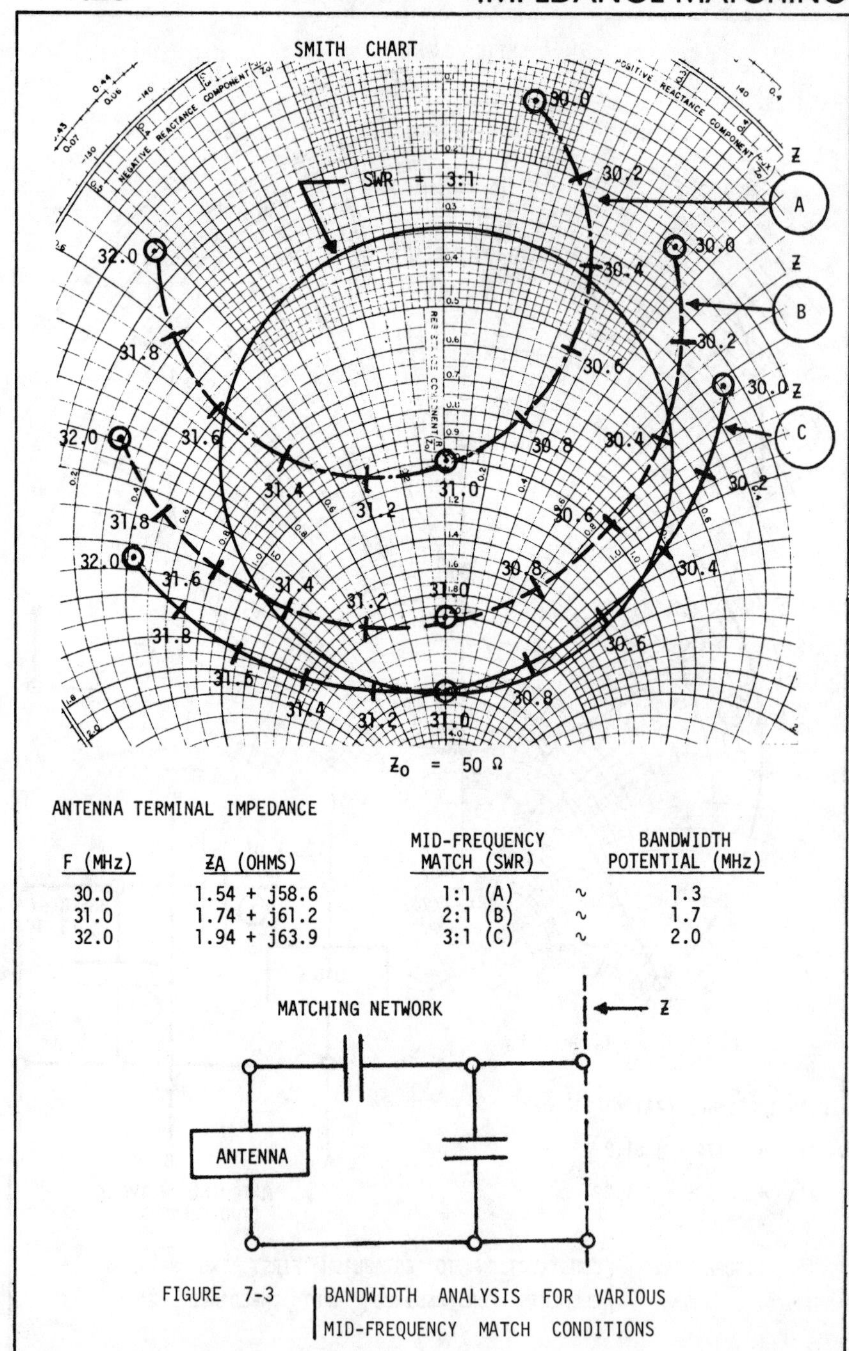

F (MHz)	Z_A (OHMS)	MID-FREQUENCY MATCH (SWR)		BANDWIDTH POTENTIAL (MHz)
30.0	1.54 + j58.6	1:1 (A)	~	1.3
31.0	1.74 + j61.2	2:1 (B)	~	1.7
32.0	1.94 + j63.9	3:1 (C)	~	2.0

FIGURE 7-3 | BANDWIDTH ANALYSIS FOR VARIOUS MID-FREQUENCY MATCH CONDITIONS

7.3 NETWORK CONFIGURATIONS

The basic four-element matching networks are shown in Figure 7-4. The networks are designated as Type I or Type II networks. Selection of the network type is dependent upon the network configuration of the first two elements. For example, should the first two elements be a shunt-series configuration, the remaining two elements form a shunt resonance circuit (Type I Network). It also follows that a series resonance circuit (Type II Network) is required when the first two elements are arranged in a series-shunt configuration.

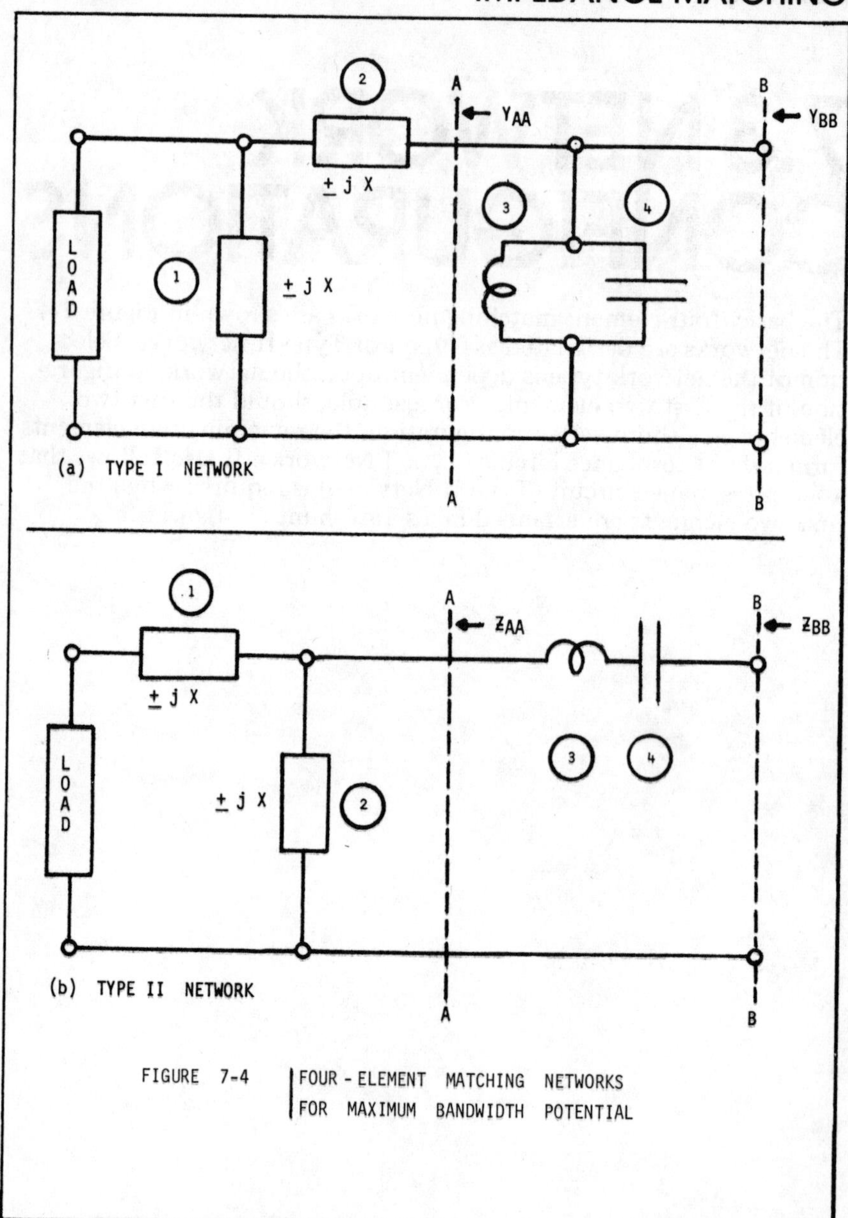

FIGURE 7-4 | FOUR - ELEMENT MATCHING NETWORKS FOR MAXIMUM BANDWIDTH POTENTIAL

7.4 NETWORK APPLICATION TO HIGH "Q" ANTENNA

The theory of broadband matching is particularly applicable for the compensation of the terminal impedance properties of high "Q" r.f. structures such as the aircraft VHF antenna that is shown in Figure 7-5 and whose impedance characteristics are given in Figure 7-6. This particular antenna is an "electrically small" open-ended slot that is cut into the leading edge of the aircraft vertical stabilizer. The structure is termed a Nitch Antenna since the length of the slot remains less than a quarter wave length throughout the 30 MHz to 76 MHz frequency band of operation.

Antennas, such as that shown in Figure 7-5, require a matching coupler for the efficient transmission of r.f. energy. It is not practical to compensate the associated high "Q" impedance characteristics, shown in Figure 7-6, with a single fixed-tuned, unless a significant sacrifice in efficiency is accepted. Multiple fixed-tuned networks are required to compensate the entire impedance curve. The more simple and least number of networks used, however, the more efficient and less complex will be the resultant coupler design. This has the implication that each fixed-tuned network, in a given design should be selected based upon maximum impedance bandwidth potential.

To illustrate the application of the broadband matching theory, the impedance properties shown in Figure 7-6 are analyzed for bandwidth potential and the selection of an optimum fixed-tuned network. Normally, such an analysis would consider various portions of the band to ascertain the number of fixed-tuned networks required to compensate the entire impedance curve. For the purpose of this

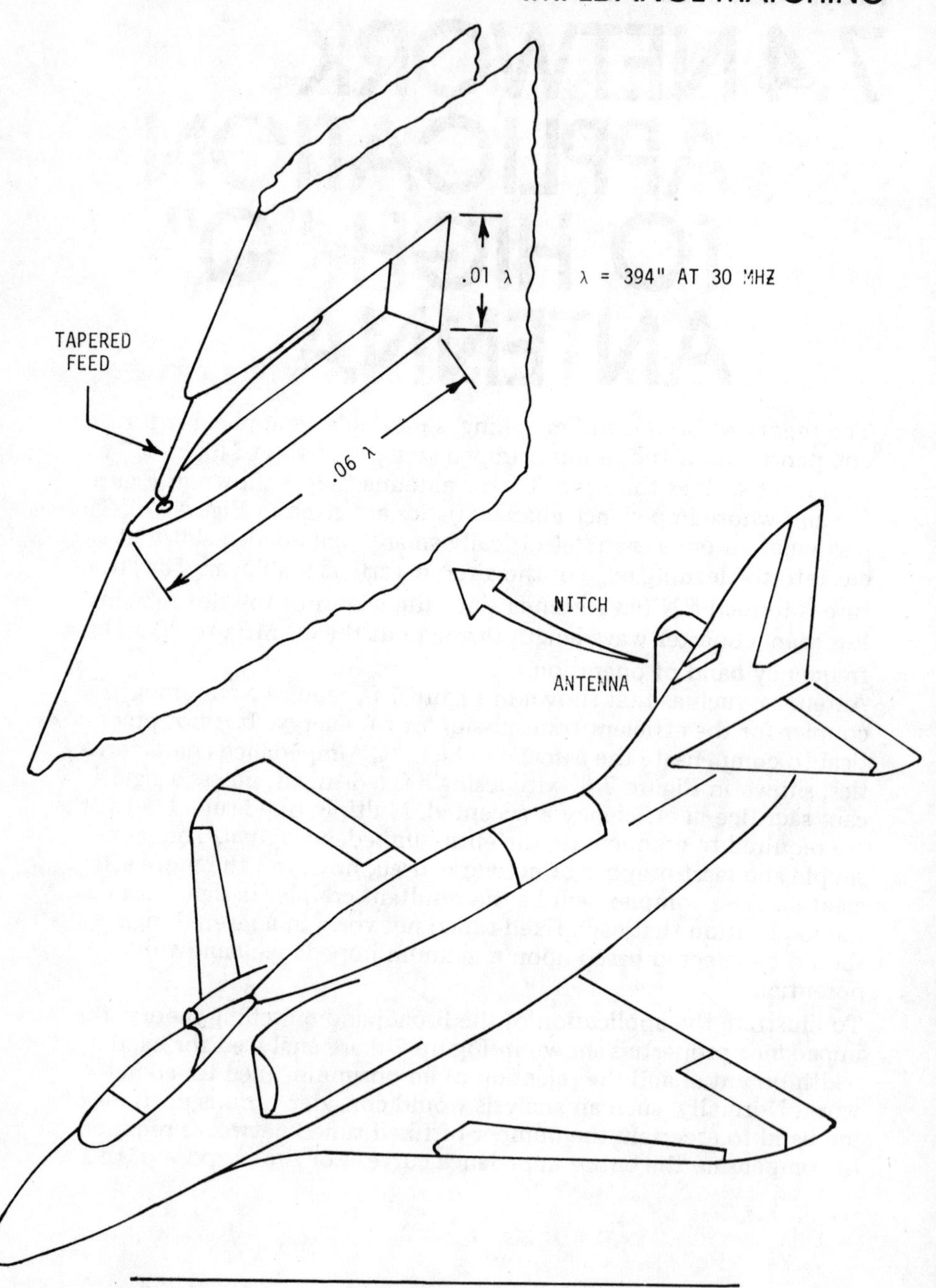

FIGURE 7-5 A TYPICAL HIGH "Q" AIRCRAFT ANTENNA

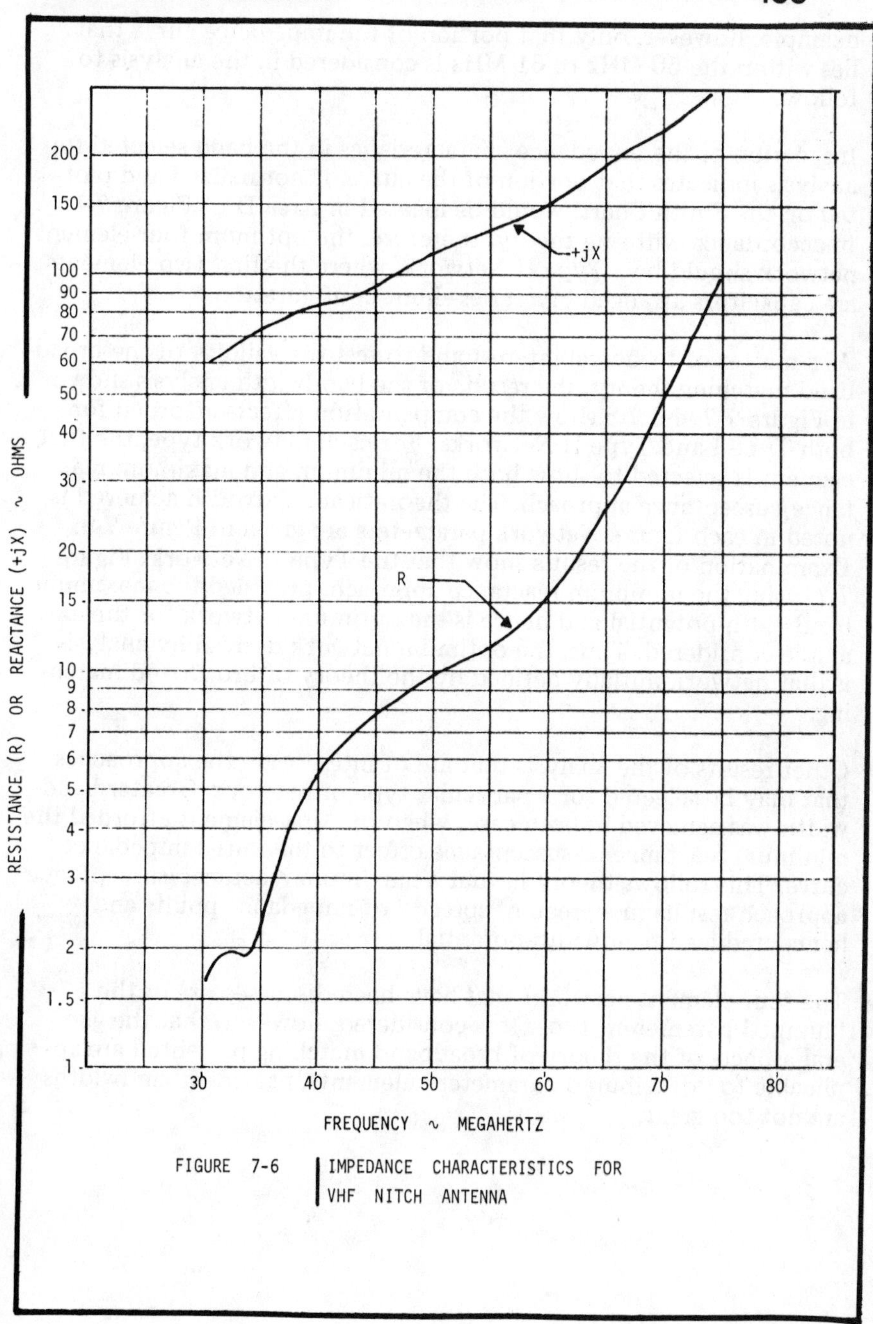

FIGURE 7-6 IMPEDANCE CHARACTERISTICS FOR VHF NITCH ANTENNA

example, however, only that portion of the impedance curve that lies within the 50 MHz to 61 MHz is considered in the analysis to follow.

Inspection of the impedance characteristics in the band selected for analysis indicates that portion of the curve, if normalized and plotted on the Smith Chart, would be located in Area D of Figure 7-1. In accordance with the theory, therefore, the optimum four-element network should be a Type II Network, where the first two elements are capacitors arranged in a series-shunt configuration.

As a matter of technical interest and to test the validity of the broadband matching theory, the results of the bandwidth analysis shown in Figure 7-7 and 7-8 show the compensation effects afforded for both Type I and Type II Networks. For each network type, the first element is selected to show both the minimum and maximum reactance/susceptance approach. The theoretical bandwidth achieved is noted in each figure. Network parameters are given in Figure 7-9. Examination of the results show that the Type II Network, Figure 7-7, using the minimum reactance approach, provided the maximum bandwidth potential and hence is the optimum network for the example considered. Thus, the optimum network derived by analysis is that network initially defined by the theory of broadband matching.

Other results of the analysis that are of interest are the approaches that may be selected for a particular type of network. Greater bandwidth was achieved in both cases when the first element afforded the minimum reactance or susceptance effect to the initial impedance curve. This follows theory in that a maximum reactance/susceptance approach results in a greater "spread" of impedance points and, hence, reduced bandwidth potential.

The four-element networks that have been discussed are of the "lumped parameter" type. It is considered, however, that the general aspects of the theory of broadband matching presented are applicable to "distributed parameter" elements, provided bandwidths are not too great.

SMITH CHART

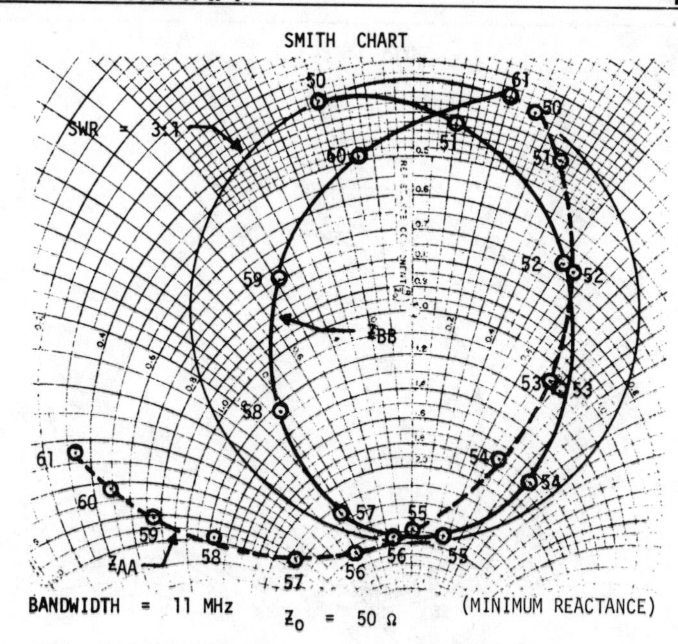

BANDWIDTH = 11 MHz (MINIMUM REACTANCE)

Z_0 = 50 Ω

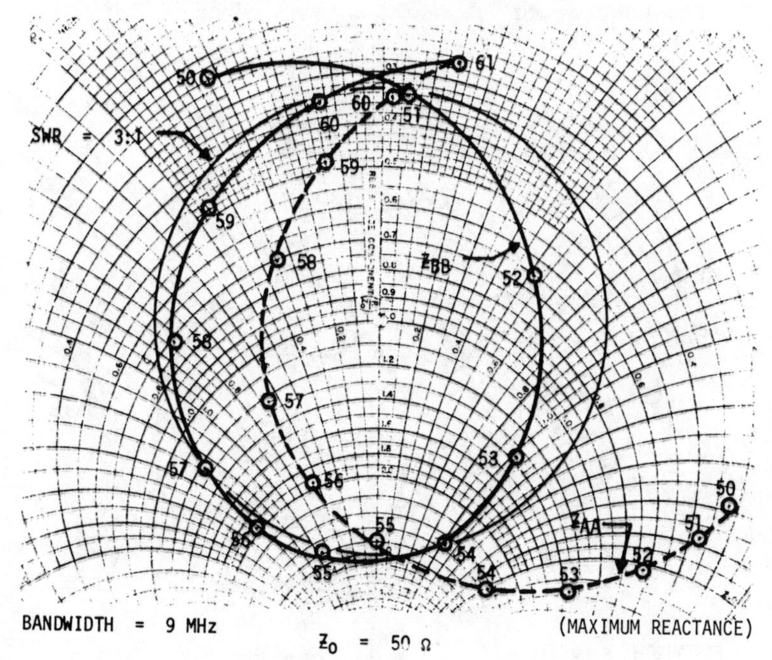

BANDWIDTH = 9 MHz (MAXIMUM REACTANCE)

Z_0 = 50 Ω

FIGURE 7-7 BANDWIDTH ANALYSIS (50-61 MHz) MAXIMUM AND MINIMUM REACTANCE APPROACHES

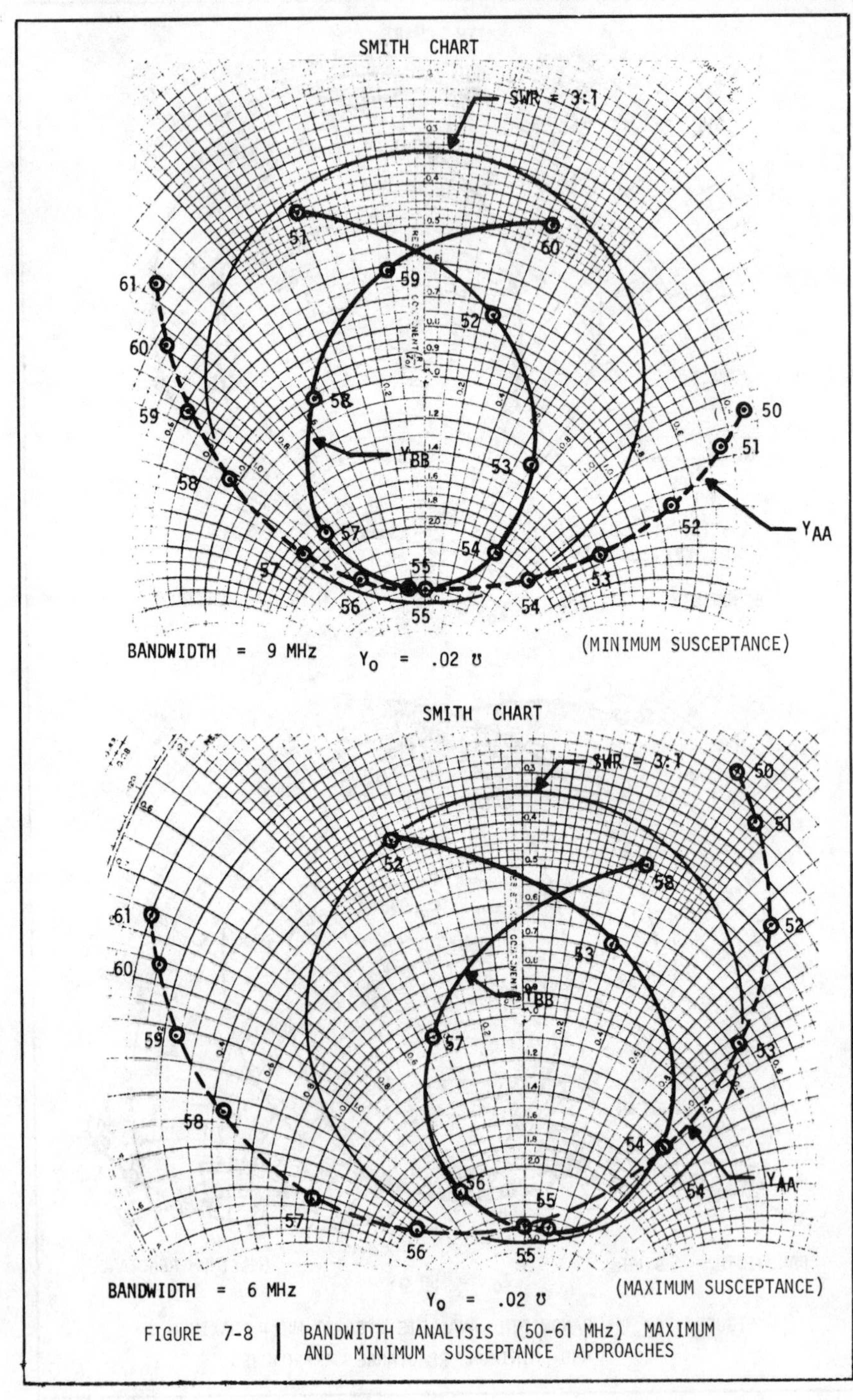

FIGURE 7-8 | BANDWIDTH ANALYSIS (50-61 MHz) MAXIMUM AND MINIMUM SUSCEPTANCE APPROACHES

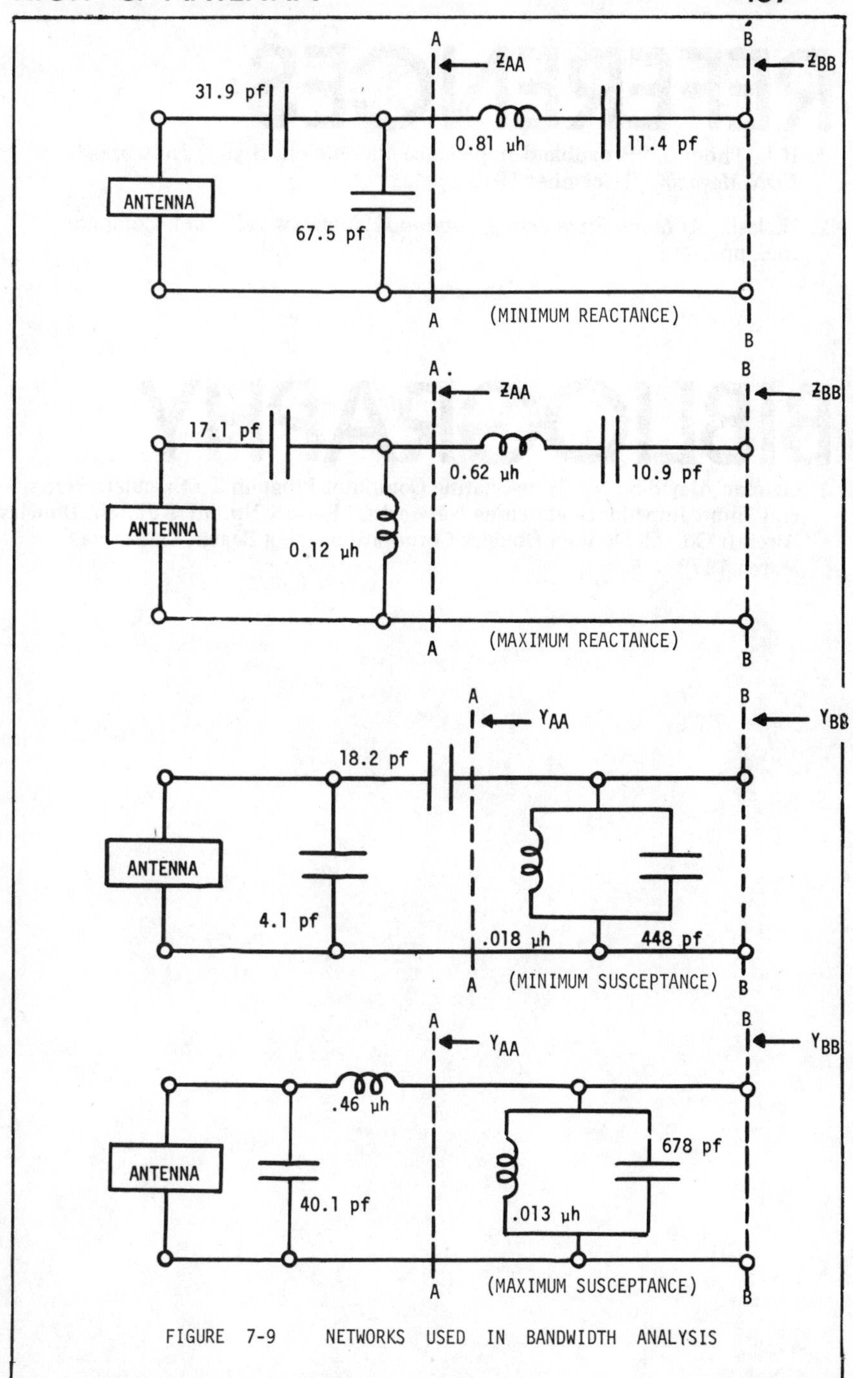

FIGURE 7-9 NETWORKS USED IN BANDWIDTH ANALYSIS

REFERENCES

1. R.L. Thomas, "Broadband Impedance Matching in High-Q Networks," *EDN Magazine*, December 1973.

2. H. Jasik, *Antenna Engineering Handbook*, McGraw-Hill Book Company, Inc., pp 2-49.

BIBLIOGRAPHY

1. Gelman, David S., "A Time-Sharing Computer Program To Calculate Series and Shunt Impedance Matching Networks," Report No. MDC J5935, Douglas Aircraft Co., McDonnell Douglas Corporation, Long Beach, California, March 1973.

8 SELECTED PRACTICAL EXAMPLES

OF IMPEDANCE MATCHING NETWORKS

8.1 INTRODUCTION

This last section presents several practical examples of impedance matching networks that utilize methods and techniques previously discussed. The examples are selected from aircraft antenna design and practice and illustrate specific matching network approaches. The selected examples are identified as follows:

Antenna Example	*Matching Network Approach*
Broadband Dipole	Single Section Series Line Transformer
VOR Slot	Single Element plus Line Positioning
UHF Blade	Two Elements arranged in a Series-Shunt Configuration

In each of the above examples, a detailed analysis of the matching approach will be given in order that the material may serve as a review for the reader.

8.2 BROADBAND DIPOLE ANTENNA

The matching network for the broadband dipole antenna illustrates the application of the single section series line transformer. The network is incorporated at the antenna terminals, as shown in Figure 8-1, for maximum compensation effectiveness. Selection of a given line transformer requires a determination of its characteristic impedance (Z_0') and its electrical length ($\theta\ell$). These two parameters completely describe the electrical characteristics of the line transformer and the degree of compensation afforded to a particular impedance curve.

The actual (Z_{AA}) and normalized $\left(Z'_{AA(50)}\right)$ terminal impedance of the dipole antenna vs. normalized frequency (f′) is given as:

f′	Z_{AA}	$Z'_{AA(50)}$
1.0	21.0 - j 5.5	0.42 - j 0.11
1.13	15.0 + j 7.5	0.30 + j 0.15
1.37	31.5 + j 14.0	0.63 + j 0.28
1.52	41.5 + j 0	0.83 + j 0
1.67	33.5 - j 21.5	0.67 - j 0.43

A schematic of the matching network is shown in Figure 8-2(a). It is desired to compensate the antenna terminal impedance curve (Z'_{AA}), shown in (b) of the figure, to an SWR of less than 2:1 at Reference Plane B-B.

8.2.1 Selecting Z_0'

An outer boundary circle is constructed in Figure 8-2(b) to enclose the impedance curve Z'_{AA} and be tangent to the SWR = 2:1 circle.

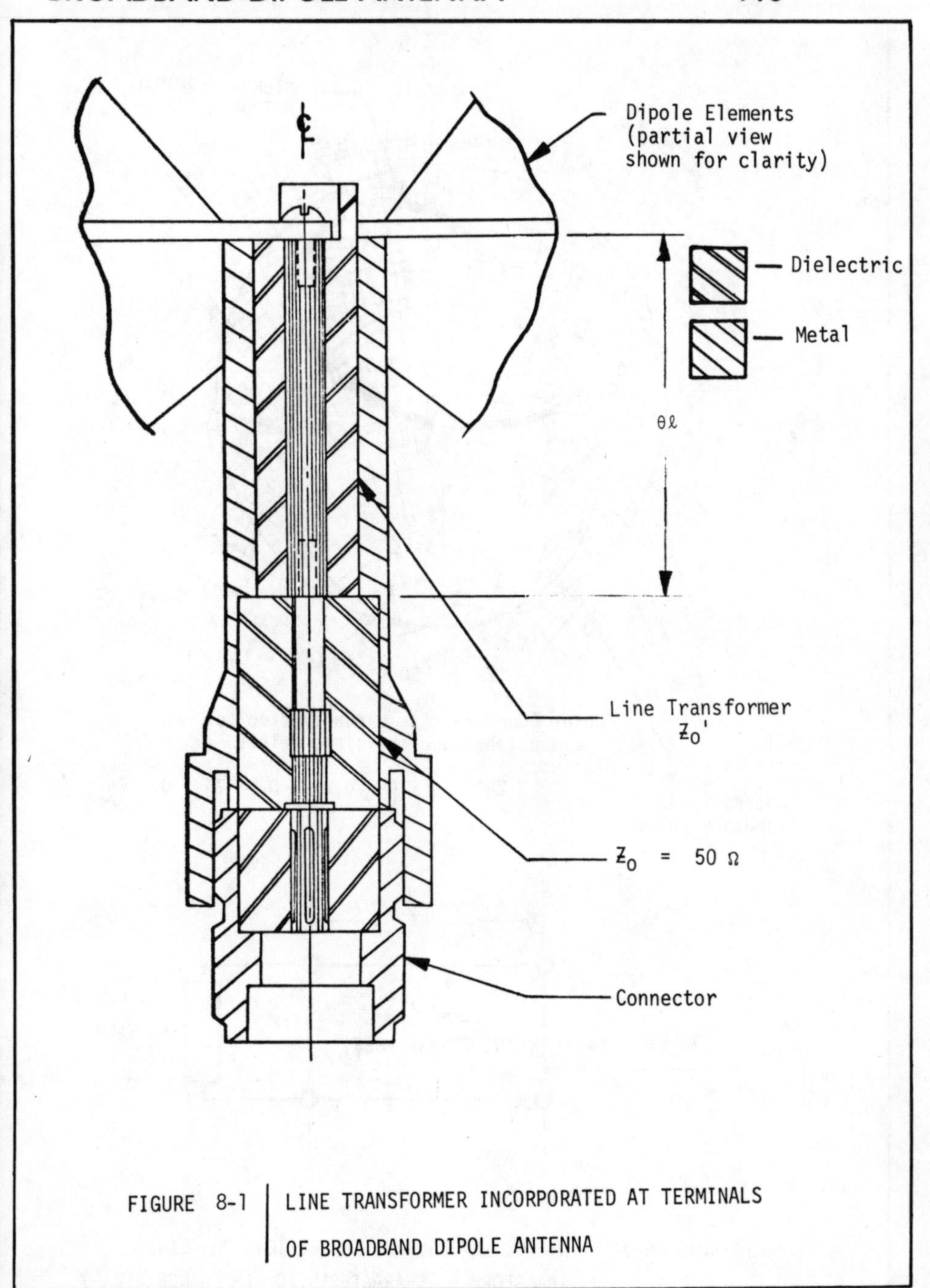

FIGURE 8-1 | LINE TRANSFORMER INCORPORATED AT TERMINALS OF BROADBAND DIPOLE ANTENNA

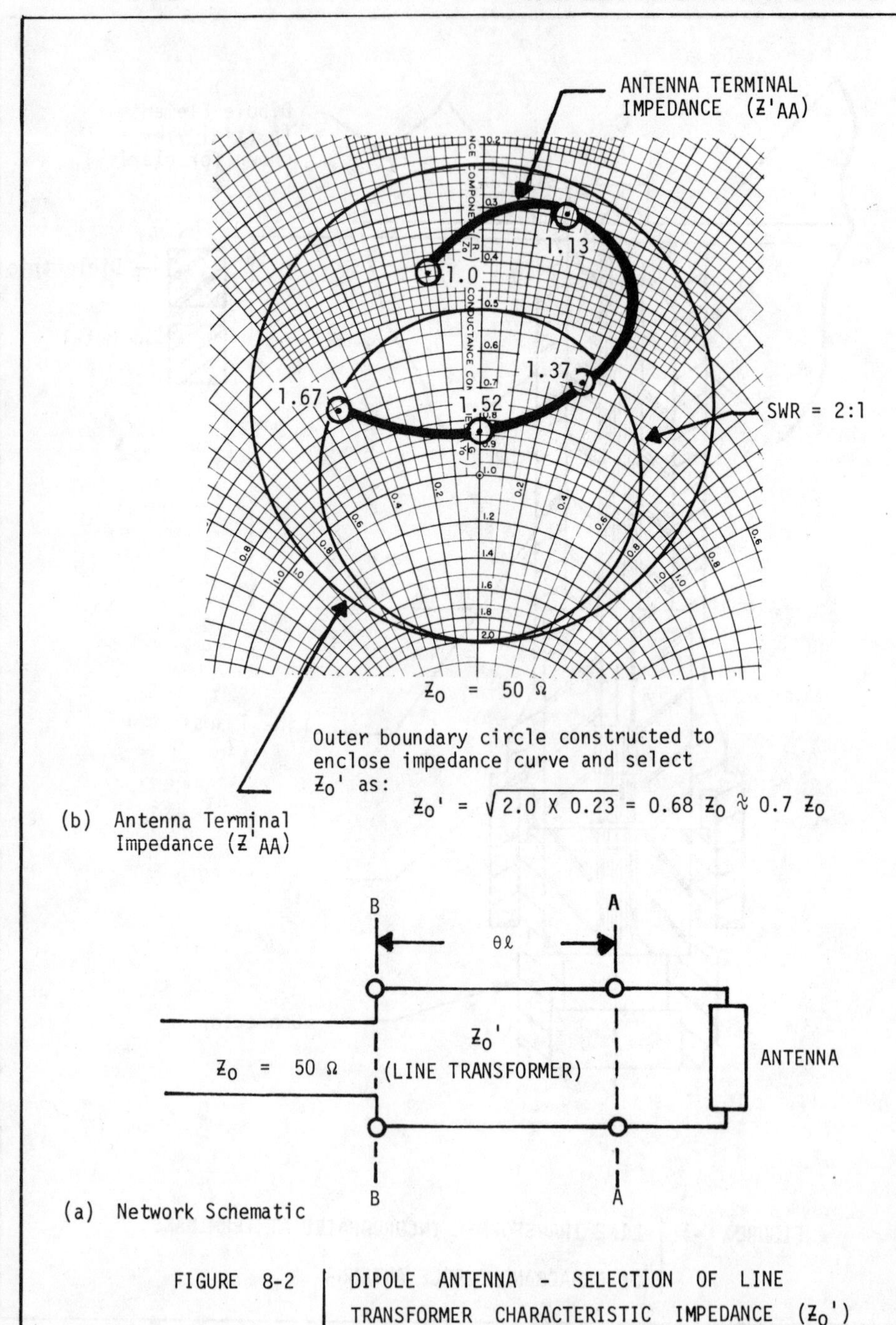

FIGURE 8-2 | DIPOLE ANTENNA - SELECTION OF LINE TRANSFORMER CHARACTERISTIC IMPEDANCE (Z_0')

This boundary circle intercepts the resistance axis at normalized values of 0.23 and 2.0. Z_0' of a possible line transformer may be found from:

$$Z_0' = \sqrt{\begin{matrix}\text{Product of Outer Boundary Circle}\\ \text{Resistance Axis Intercepts}\end{matrix}}$$

and for the example is:

$$Z_0' = \sqrt{2.0 \times 0.23} = 0.68\ Z_0$$

Since the transformer design charts given in Appendix A are in increments of 0.1 Z_0, it is best to select a Z_0' of 0.7 Z_0 as nearest to that of the above calculated value.

8.2.2 Analysis to Determine $\theta \ell$

The antenna terminal impedance (Z'_{AA}) is next plotted on the $Z_0' =$ 0.7 Z_0 design chart, shown in Figure 8-3, for an analysis to determine the transformer electrical length ($\theta \ell$). A trial value for $\theta \ell$ is best selected for that point on the impedance curve that appears most critical. Examination of the figure shows that the 1.13 impedance point is probably the most critical as it lies very near the outer boundary circle. Since this point lies within the $\theta = 75°$ circle, $\theta \ell$ is selected to be 75° at f = 1.13. The electrical length of the transformer at the other frequencies is then:

f	$\theta \ell$ (degrees)
1.0	66.4
1.13	75.0
1.37	90.9
1.52	100.9
1.67	110.9

By definition, an impedance point lying within its respective θ circle for that frequency will be transformed to within the specified SWR circle. Examination of Figure 8-3 shows that all impedance points fall within their respective θ circles, and hence the entire impedance curve, Z'_{AA}, may be transformed to within the SWR = 2:1 circle using a transformer whose $Z_0' = 0.7\ Z_0$ and $\theta \ell = 75°$ at f = 1.13. In the case shown, $Z_0 = 50$ ohms and therefore, $Z_0' = 35$ ohms.

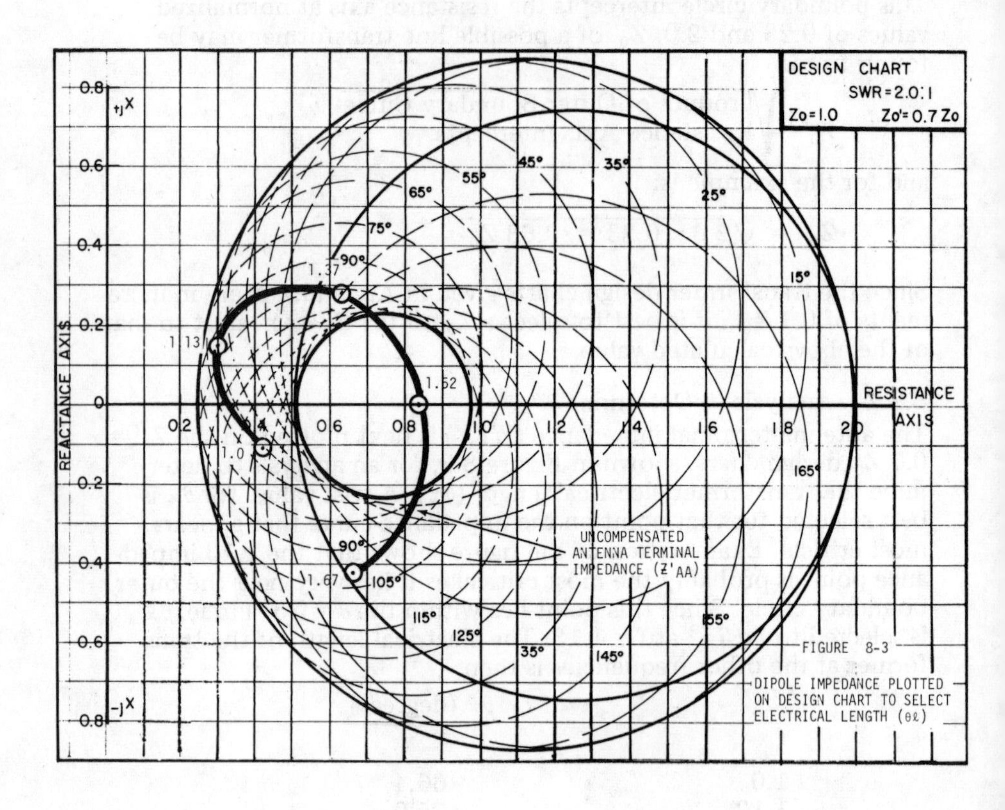

FIGURE 8-3
DIPOLE IMPEDANCE PLOTTED ON DESIGN CHART TO SELECT ELECTRICAL LENGTH (θℓ)

8.2.3 Verification Analysis

The values for Z_0' and $\theta \ell$, determined by the analysis, are verified in Figure 8-4. In (a) of the figure, the antenna impedance is normalized to $Z_0' = 35$ ohms and rotated across the transformer to Reference Plane B-B to give impedance curve $Z'_{BB(35)}$. Z'_{BB}, normalized to $Z_0 = 50$ ohms, is shown in (b) of the figure. The results are thus verified in that all impedance points are within the SWR circle.

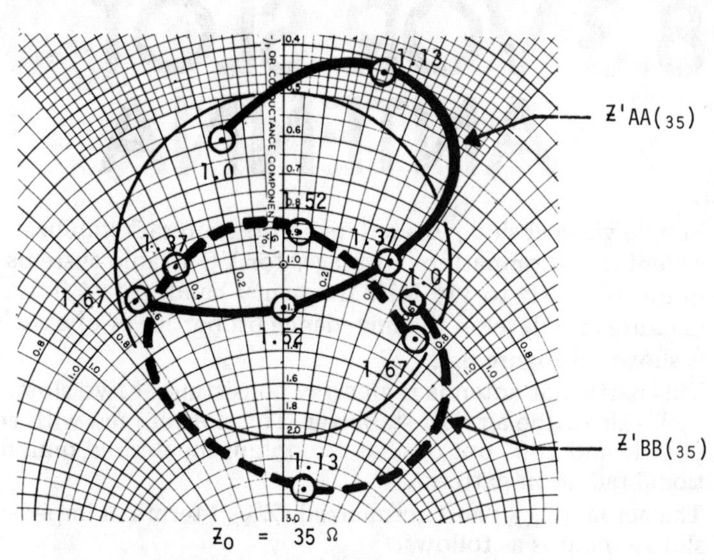

(a) Transformation of antenna impedance across line transformer where $\theta\ell = 0.184\ \lambda$ at f = 1.0

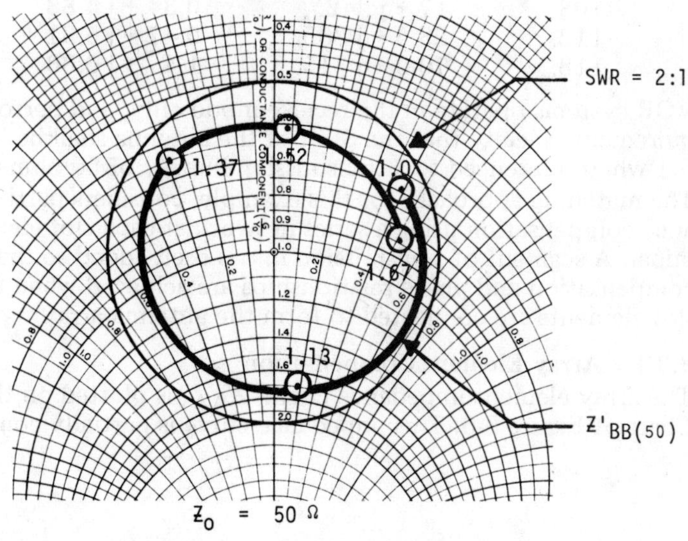

(b) Compensated Impedance Curve (Z'_{BB})

FIGURE 8-4 | DIPOLE ANTENNA - IMPEDANCE COMPENSATION USING SINGLE SECTION LINE TRANSFORMER

8.3 VOR SLOT ANTENNA

Simple vertical slots, located high in the aircraft stabilizer, provide excellent radiation pattern coverages for VOR radio systems operating in the 108 MHz to 118 MHz frequency band. A test fixture used to measure and develop the impedance properties of one such antenna is shown in Figure 8-5.

This particular antenna is an array consisting of two slots: one on each side of the aircraft stabilizer. The slots are fed with equal amplitude and 180° out of phase to obtain the desired omnidirectional radiation patterns.

The actual (Z_{AA}) and normalized (Z'_{AA}) terminal impedance of each slot element is as follows:

f′ (MHz)	Z_{AA}	$Z'_{AA(50)}$
108.	17 + j 242.	0.34 + j 4.88
112.	23.5 + j 265.	0.47 + j 5.31
118.	70.0 + j 315.	1.4 + j 6.30

VOR systems operate in the receive mode only, and performance requirements specify that the SWR of the antenna shall be less than 5:1 when referenced to a transmission line Z_0 of 50 ohms.

The matching approach, for this example, considers partial impedance compensation using a series network at each slot element terminal. A segment of transmission line is then added so that final compensation and phase relationships are achieved when the two slot elements are connected to form the antenna array.

8.3.1 Array Element Compensation

The array element impedance curve, Z'_{AA}, is plotted on the Smith Chart of Figure 8-6. This impedance curve is partially compensated

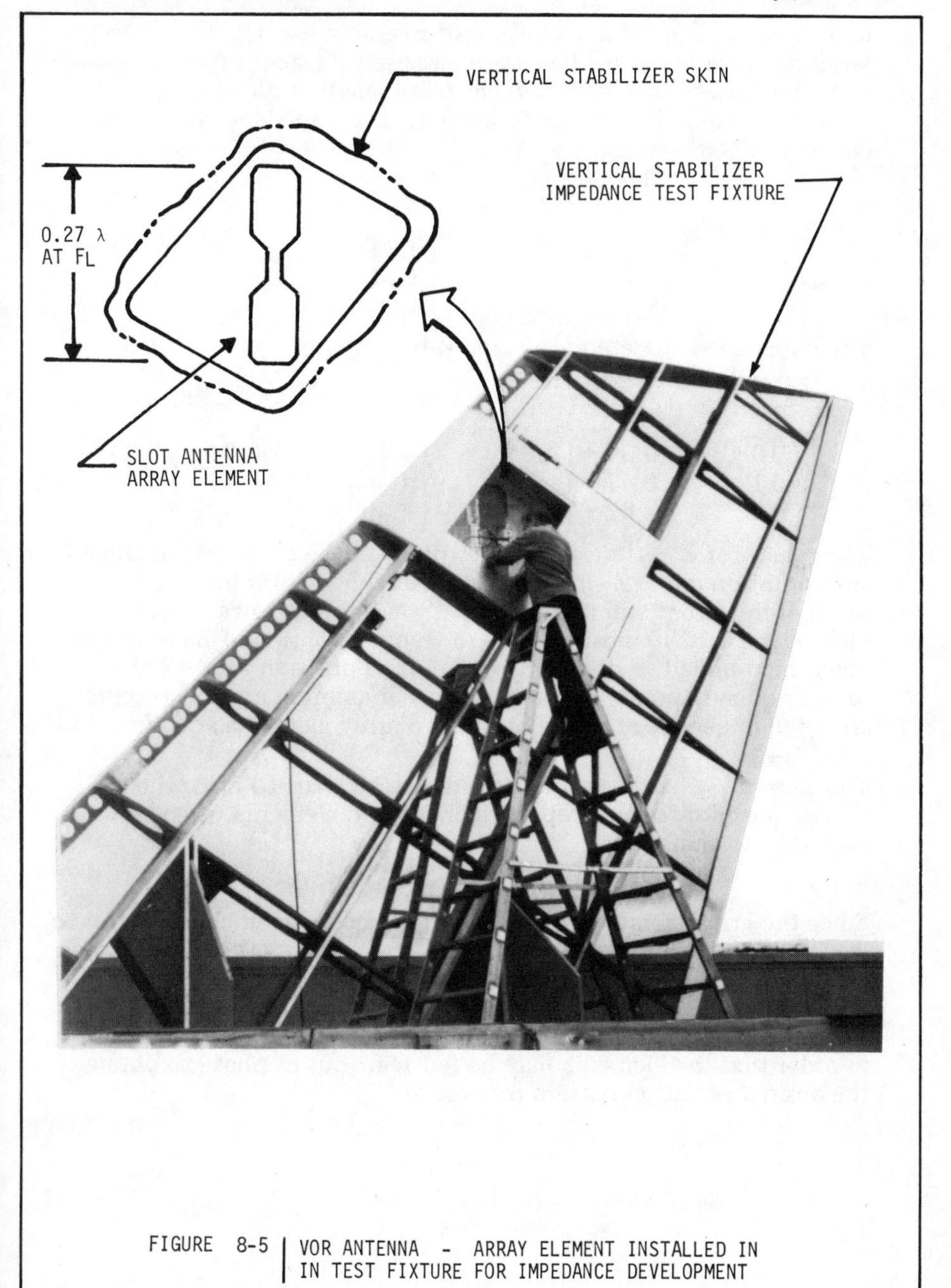

FIGURE 8-5 | VOR ANTENNA - ARRAY ELEMENT INSTALLED IN IN TEST FIXTURE FOR IMPEDANCE DEVELOPMENT

to give curve Z'_{BB} by the addition of a series network, $Z'_{①}$. The series network consists of an open circuited segment of transmission line whose Z_0 = 100 ohms and electrical length = .067 λ at f = 112 MHz. This segment adds the following series impedance to the array element impedance:

f (MHz)	$Z'_{①}$
108	0 - j 4.64
112	0 - j 4.44
118	0 - j 4.20

The impedance at Reference Plane B-B is then:

f (MHz)	Z'_{AA}	+	$Z'_{①}$	=	Z'_{BB}
108	0.34 + j 4.88		- j 4.64		0.34 + j 0.24
112	0.47 + j 5.31		- j 4.44		0.47 + j 0.87
122	1.42 + j 6.30		- j 4.20		1.40 + j 2.1

The values for $Z'_{①}$ are selected to position the f = 112 MHz impedance point on curve Z'_{BB} such that a radial line from the chart center through this point will intersect the "wavelengths towards generator" scale at the 0.125 λ position. A λ/8 segment, $Z'_{②}$, of Z_0 = 50 ohm line is next added to obtain curve Z'_{CC} and position the 112 MHz (mid-frequency point) on the resistance maximum axis. Z'_{CC} represents the impedance of each of the two array elements at Reference Plane C-C.

The curve Z'_{CC} has been 'positioned' on the chart to favor the addition of the admittance properties of the two elements when connected to form the array.

8.3.2

Since the array elements are to be connected in parallel it is better to work with admittances. Figure 8-7 describes the method used to connect the array elements. In the figure, Y'_{CC} = Y'_{FF} since the matching networks ① and ② are identical for each array element. A λ/2 segment of Z_0 = 50 ohms line is added at Reference Plane F-F in order that the elements may be fed 180° out of phase to obtain the desired radiation pattern properties.

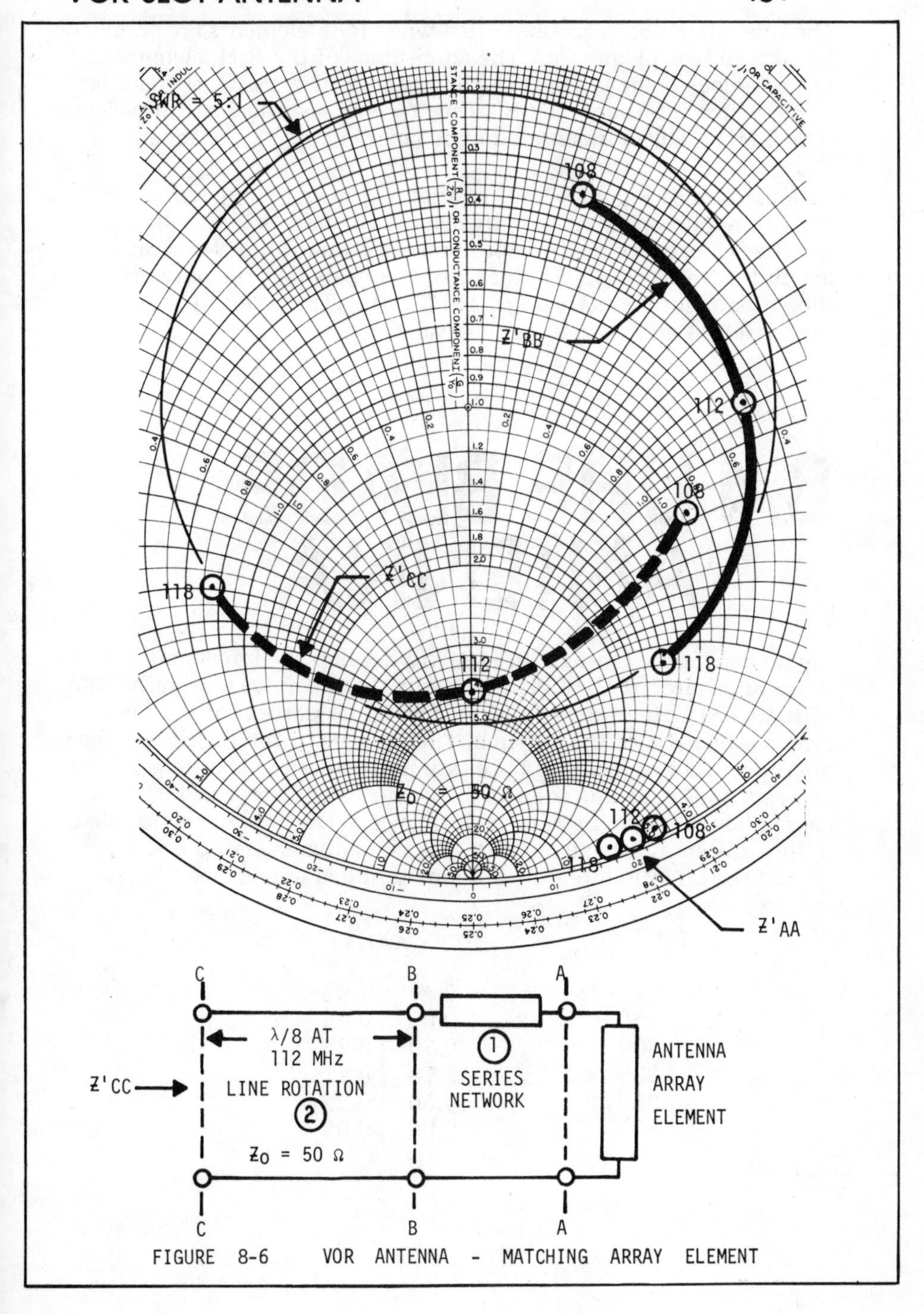

FIGURE 8-6 VOR ANTENNA - MATCHING ARRAY ELEMENT

The final matching is accomplished when both elements are connected at Reference Plane C-C. The admittance of the R.H. element at this reference plane is Y'_{CC}. The admittance of the L.H. element is Y''_{CC}, which is obtained by rotating Y'_{FF} an electrical half wavelength at 112 MHz. The compensated impedance curve, Y'_M, is equal to $Y'_{CC} + Y''_{CC}$.

Examination of Figure 8-7 shows that Y'_M meets the specified SWR $=< 5:1$ requirement but is marginal at 118 MHz. A better match may be achieved by a more optimum selection of the electrical length for Network ②. In this case, it appears that the network should be slightly less than $\lambda/8$ at 112 MHz. The proof of this, however, is left as an exercise for the reader.

8.4 UHF BLADE ANTENNA

Figure 8-8 shows a test fixture used to develop the impedance properties of an aircraft UHF blade antenna that operates within the 225 MHz to 400 MHz frequency band. The upper portion of the figure gives a schematic of the antenna. The purpose of the r.f. chokes that are shown is to suppress the excitation of undesired traveling wave currents on the leading edge of the vertical stabilizer. Such currents cause poor radiation pattern performance.[1] The antenna is installed in the tail-cap of a small aircraft. The measurement setup is similar to that shown in Figure 4-2. Antenna terminal impedance $Z'_{AA(50)}$ is:

f (MHz)	$Z'_{AA(50)}$
225	0.45 - j 1.02
260	0.50 - j 0.65
300	0.70 - j 0.32
335	1.18 - j 0.21
350	0.56 - j 0.23
400	0.90 - j 0.05

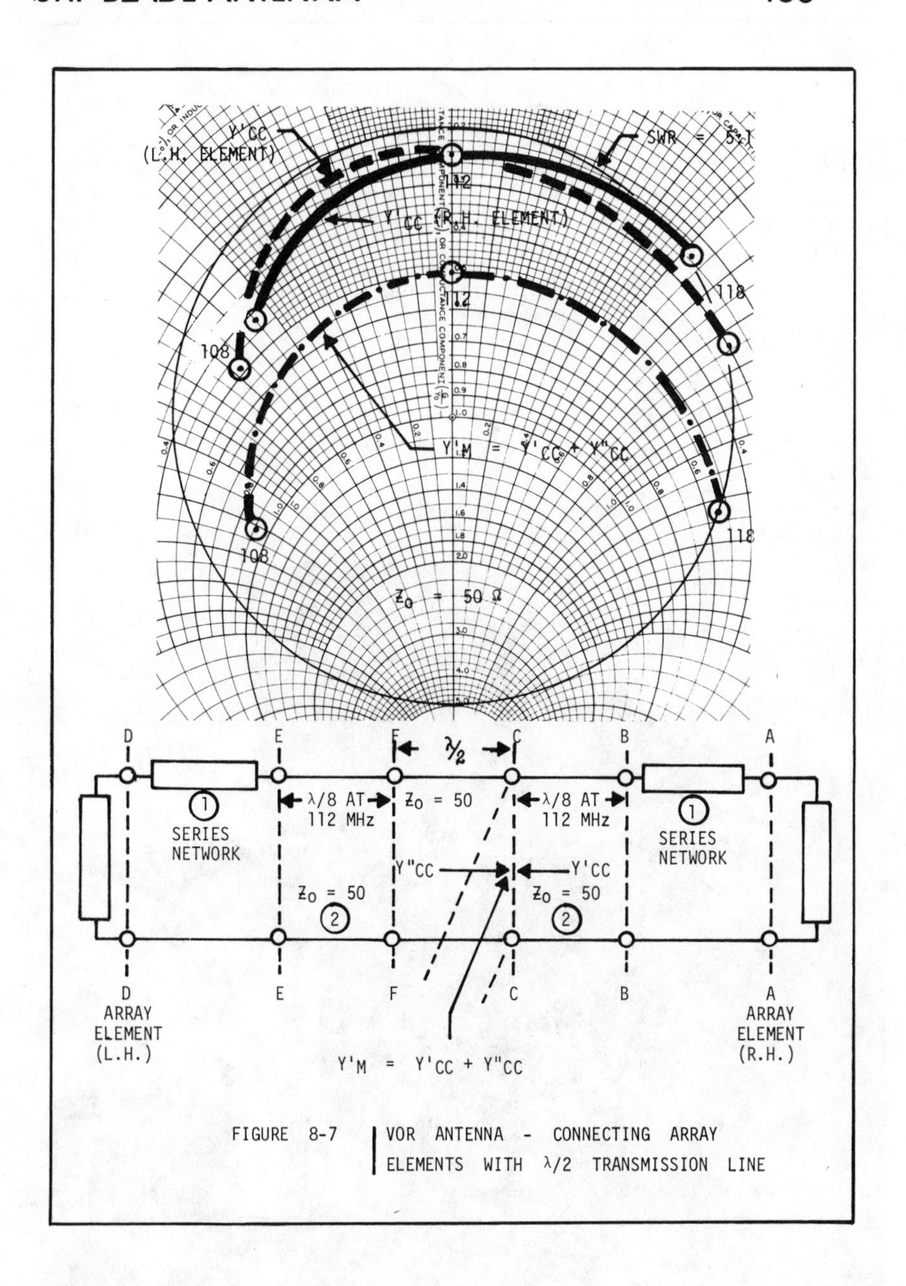

FIGURE 8-7 | VOR ANTENNA - CONNECTING ARRAY ELEMENTS WITH λ/2 TRANSMISSION LINE

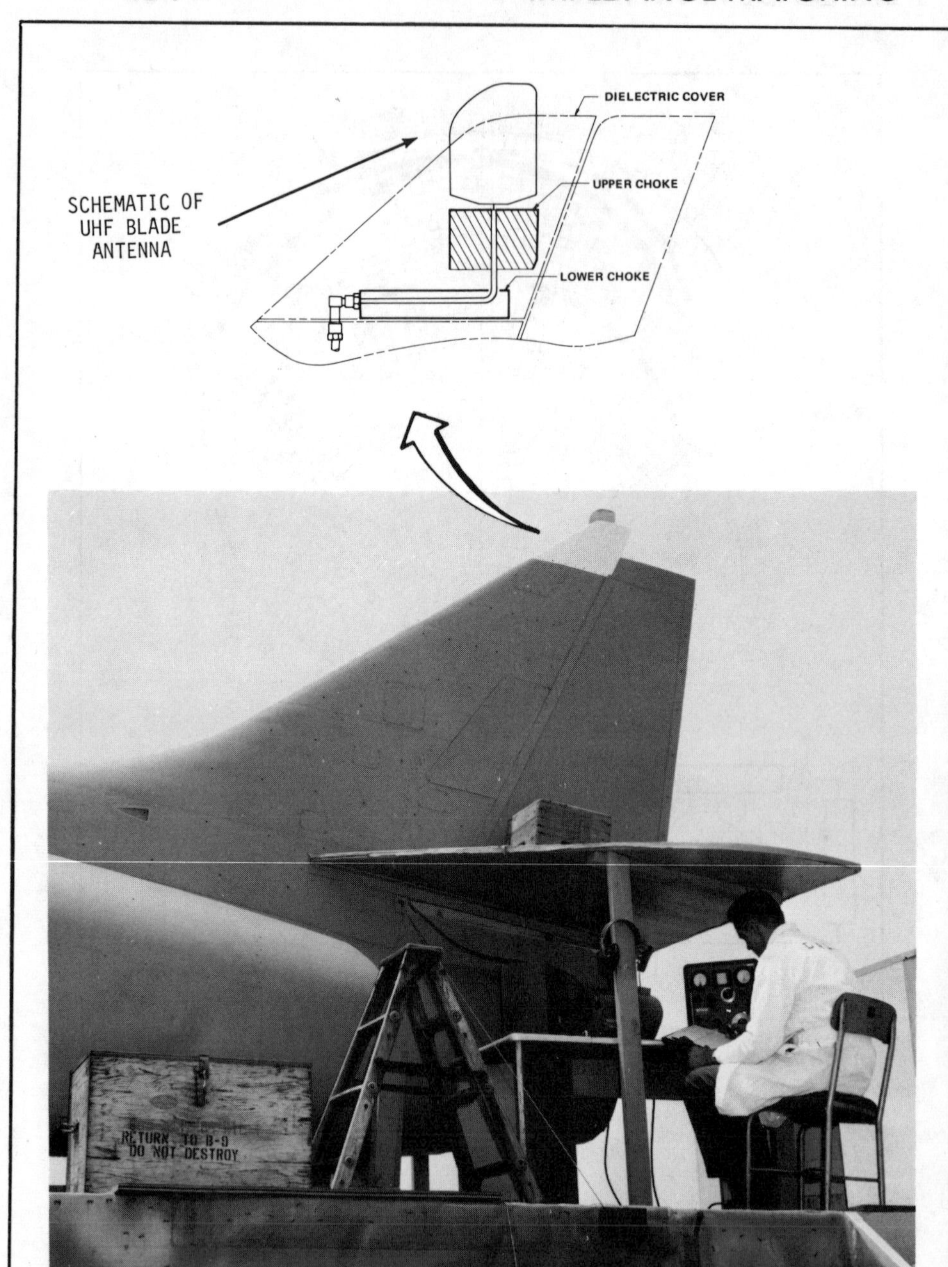

FIGURE 8-8 | TEST FIXTURE USED TO DEVELOP IMPEDANCE PROPERTIES OF UHF BLADE

The antenna is designed to operate in both the transmit and receive modes and, as such, the SWR must be less than 2:1 referenced to a 50 ohm transmission line. A two-element network arranged in a series-shunt configuration is selected to compensate the antenna impedance.

8.4.1 First Matching Element

Z'_{AA} is plotted in Figure 8-9(a). Examination of the curve shows that partial compensation may be obtained by a series inductive network. This network should be selected such that the end frequency points are "balanced" about the resistance axis and will be susceptible to compensation by the second network. A series short-circuited segment of 50 ohm line (less than $\lambda/4$) will provide inductive reactance. An electrical length ($\theta\ell$) of .059 λ at 225 MHz is selected as a trial value. The impedance, $Z'_{①}$, added by this segment over the frequency band is:

f (MHz)	$\theta\ell_{①}$ (λ)	$Z'_{①}$ (50)
225	.059	+ j 0.39
260	.068	+ j 0.46
300	.079	+ j 0.54
335	.088	+ j 0.62
350	.092	+ j 0.65
400	.105	+ j 0.78

And the impedance, Z'_{BB}, at Reference Plane B-B follows as:

f (MHz)	Z'_{AA}	+	$Z'_{①}$	=	Z'_{BB}
225	0.45 - j 1.02		+ j 0.39		0.45 - j 0.63
260	0.50 - j 0.65		+ j 0.46		0.50 - j 0.19
300	0.70 - j 0.32		+ j 0.54		0.70 + j 0.22
335	1.18 - j 0.21		+ j 0.62		1.18 + j 0.41
350	0.56 - j 0.23		+ j 0.65		0.56 + j 0.42
400	0.90 - j 0.05		+ j 0.78		0.90 + j 0.73

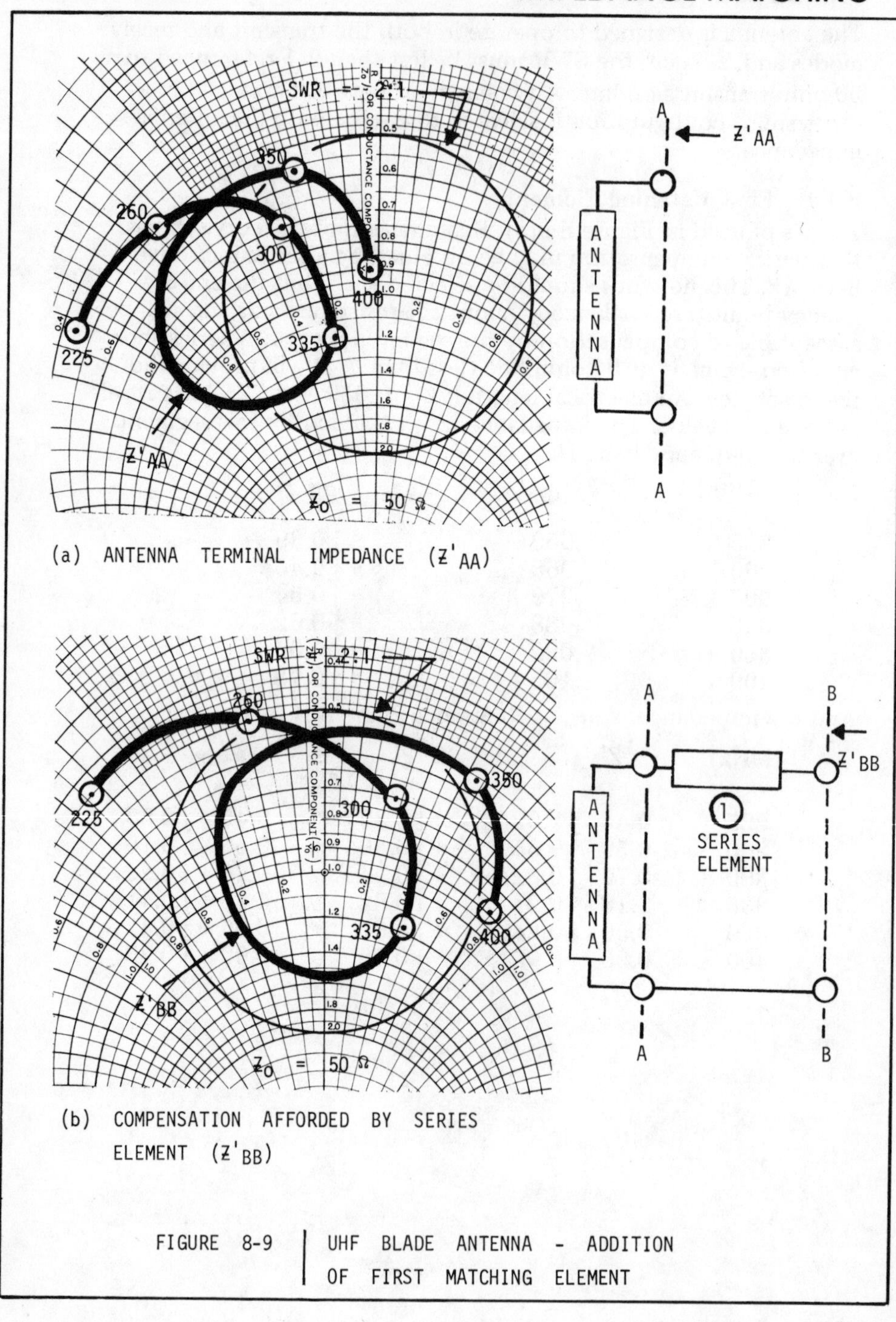

(a) ANTENNA TERMINAL IMPEDANCE (Z'AA)

(b) COMPENSATION AFFORDED BY SERIES ELEMENT (Z'BB)

FIGURE 8-9 | UHF BLADE ANTENNA - ADDITION OF FIRST MATCHING ELEMENT

Z'_{BB} is plotted in Figure 8-9(b) and Y'_{BB} is shown in Figure 8-10(a). The trial value selected for $\theta \ell$ appears to be reasonable as inspection of Y'_{BB} curve shows that the end frequency points will be susceptible to compensation by the second network which, in this case, is a shunt network.

8.4.2 Second Matching Network

The values of Y'_{BB} from Figure 8-10(a) are:

f (MHz)	$Y'_{BB(.02)}$
225	0.75 + j 1.04
260	1.75 + j 0.65
300	1.30 - j 0.40
335	0.75 - j 0.26
350	1.13 - j 0.85
400	0.67 - j 0.55

The approach for the second network is to "wrap" the end frequency points to within the SWR = 2:1 circle. The admittance at 400 MHz, Y'_{BB} = 0.67 - j 0.55, is examined to determine a trial selection for the second network. If the second network adds an admittance of $Y'_{②}$ = + j 0.99 at 400 MHz, the admittance at Reference Plane C-C will be:

$$
\begin{aligned}
Y'_{CC} &= Y'_{BB} + Y'_{2} \\
&= 0.67 - j\,0.55 + j\,0.99 \\
&= 0.67 + j\,0.44
\end{aligned}
$$

And the 400 MHz impedance point is positioned to be within the SWR = 2:1 circle. A segment of open-circuited 50 ohm line, $\theta \ell_{2}$ = 0.624 λ at 400 MHz, will provide a normalized shunt admittance of + j 0.99. The admittance of this segment and the resultant admittance at Reference Plane C-C over the frequency band is:

f (MHz)	$\theta \ell_{②}$ (λ)	$Y'_{②}$	+	Y'_{BB}	=	Y'_{CC}
225	0.351	- j 1.37		0.75 + j 1.04		0.75 - j 0.33
260	0.405	- j 0.68		1.75 + j 0.65		1.75 - j 0.03
300	0.467	- j 0.21		1.30 - j 0.40		1.30 - j 0.61
335	0.522	+ j 0.14		0.75 - j 0.26		0.75 - j 0.12
350	0.545	+ j 0.29		1.13 - j 0.85		1.13 - j 0.56
400	0.624	+ j 0.99		0.67 - j 0.55		0.67 + j 0.44

Y'_{CC} is plotted in (b) of Figure 8-10. The impedance curve falls within the specified SWR circle. The trial values selected for matching networks ① and ② are therefore satisfactory.

8.5 CONCLUDING REMARKS

The material presented in this work provides the reader with a practical knowledge of impedance matching methods and techniques. Both simple and complex networks have been discussed in light of narrow and broadband matching requirements. Emphasis has been placed upon the use of graphical aids in the solution of network problems, as such aids provide the reader with a visual interpretation of parameter variation. Proficiency in selecting impedance matching networks must, as in any other discipline, follow from practice.

REFERENCE

1. A.R. Ellis, "UHF Tail-cap Antenna Pattern Characteristics and Their Control," Stanford Research Institute Technical Report 35, S.R.I, Project 1197, February 1955.

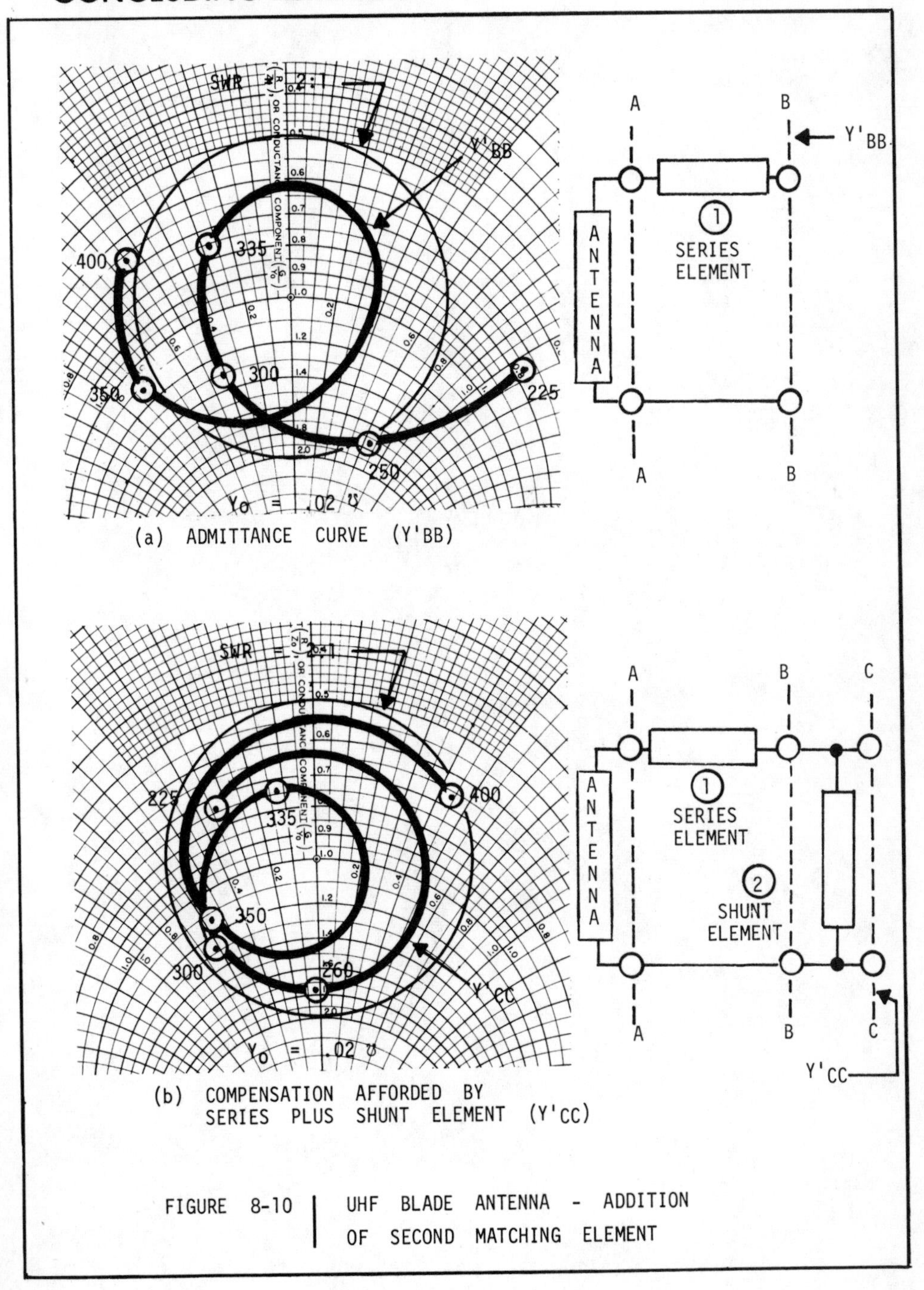

FIGURE 8-10 | UHF BLADE ANTENNA - ADDITION OF SECOND MATCHING ELEMENT

Appendix A

DESIGN CHARTS FOR SINGLE SECTION LINE TRANSFORMERS

Definition Circle = 2:1

$Z_0 = 1.0$

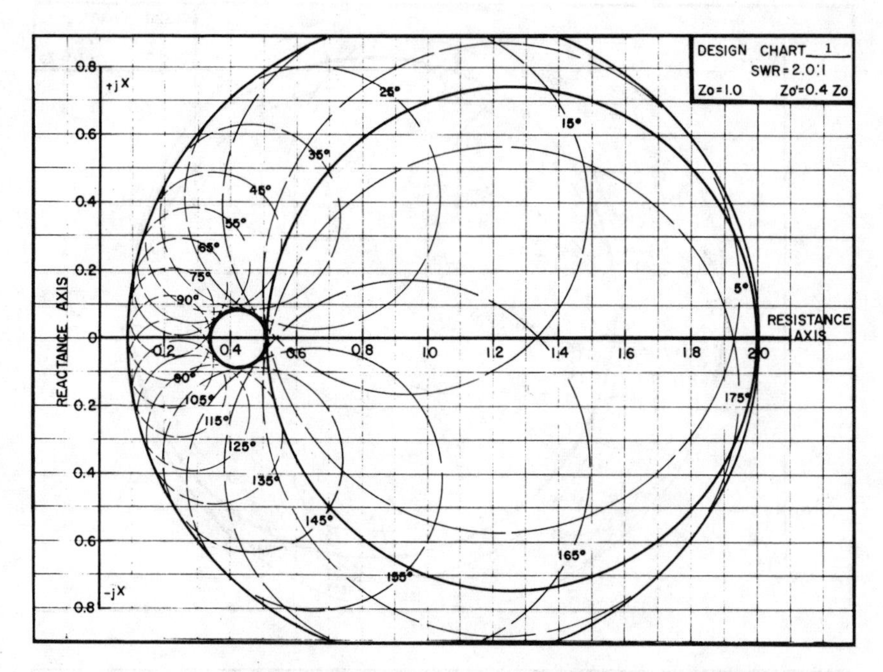
DESIGN CHART 1
SWR = 2.0:1
Zo = 1.0
Zo' = 0.4 Zo
+jX
-jX
REACTANCE AXIS
RESISTANCE AXIS
0.8
0.6
0.4
0.2
0
0.2
0.4
0.6
0.8
1.0
1.2
1.4
1.6
1.8
2.0
5°
15°
25°
35°
45°
55°
65°
75°
90°
105°
115°
125°
135°
145°
155°
165°
175°

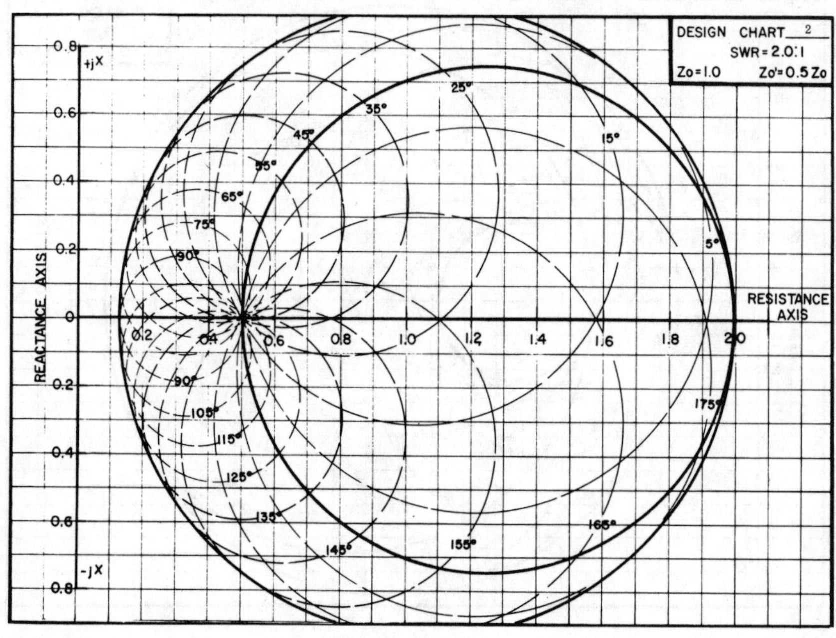
DESIGN CHART 2
SWR = 2.0:1
Zo = 1.0
Zo' = 0.5 Zo
+jX
-jX
REACTANCE AXIS
RESISTANCE AXIS
0.8
0.6
0.4
0.2
0
0.2
0.4
0.6
0.8
1.0
1.2
1.4
1.6
1.8
2.0
5°
15°
25°
35°
45°
55°
65°
75°
90°
105°
115°
125°
135°
145°
155°
165°
175°

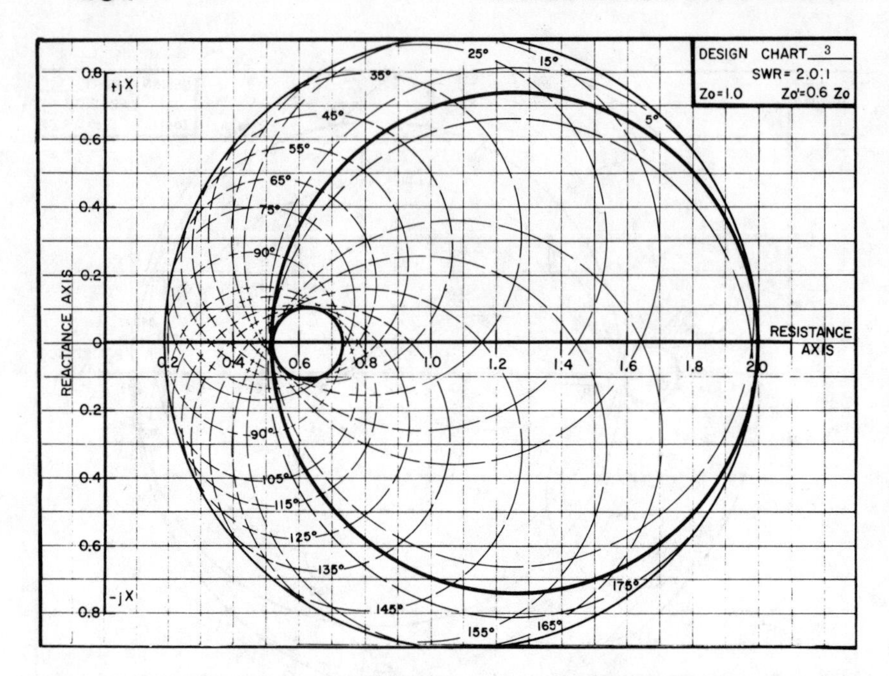
DESIGN CHART 3
SWR = 2.0:1
Zo = 1.0
Zo' = 0.6 Zo
+jX
-jX
REACTANCE AXIS
RESISTANCE AXIS
0.8
0.6
0.4
0.2
0
0.2
0.4
0.6
0.8
1.0
1.2
1.4
1.6
1.8
2.0
5°
15°
25°
35°
45°
55°
65°
75°
90°
105°
115°
125°
135°
145°
155°
165°
175°

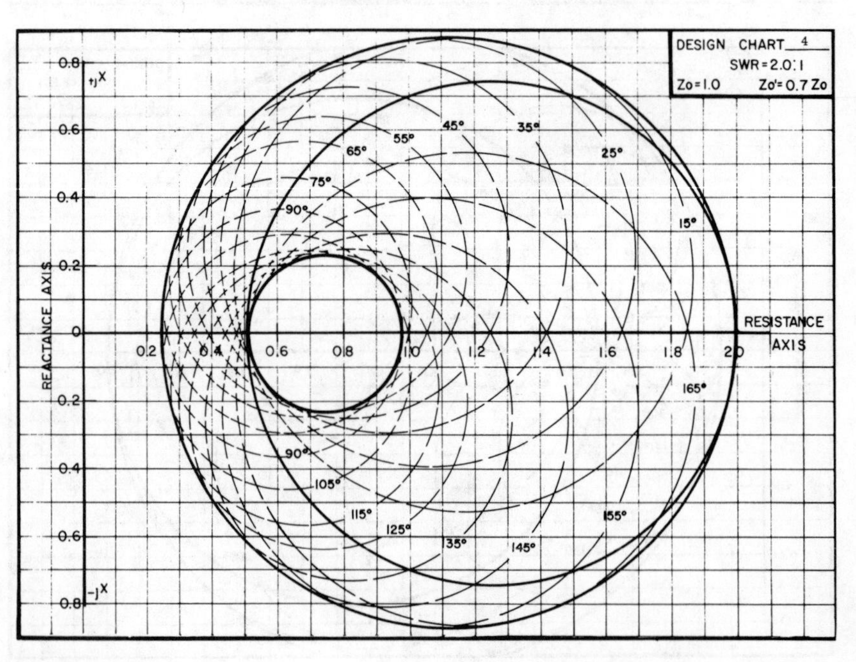
DESIGN CHART 4
SWR = 2.0:1
Zo = 1.0
Zo' = 0.7 Zo
+jX
-jX
REACTANCE AXIS
RESISTANCE AXIS
0.8
0.6
0.4
0.2
0
0.2
0.4
0.6
0.8
1.0
1.2
1.4
1.6
1.8
2.0
15°
25°
35°
45°
55°
65°
75°
90°
105°
115°
125°
135°
145°
155°
165°

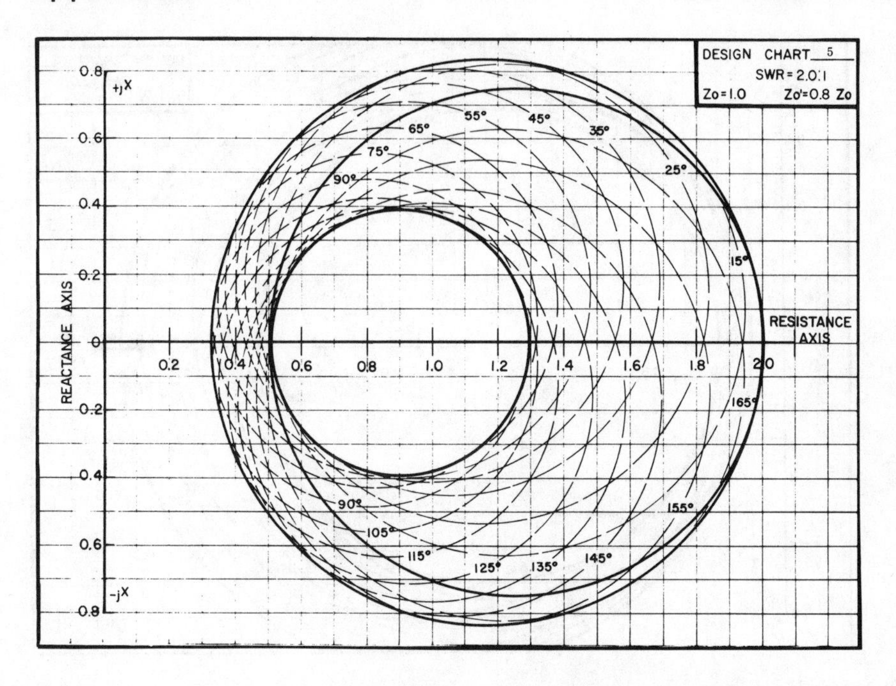
DESIGN CHART 5
SWR = 2.0:1
Zo = 1.0 Zo' = 0.8 Zo
REACTANCE AXIS
RESISTANCE AXIS
+jX
-jX
0.8
0.6
0.4
0.2
0
0.2
0.4
0.6
0.8
0.2
0.4
0.6
0.8
1.0
1.2
1.4
1.6
1.8
2.0
65°
55°
45°
35°
75°
90°
25°
15°
165°
155°
90°
105°
115°
125°
135°
145°

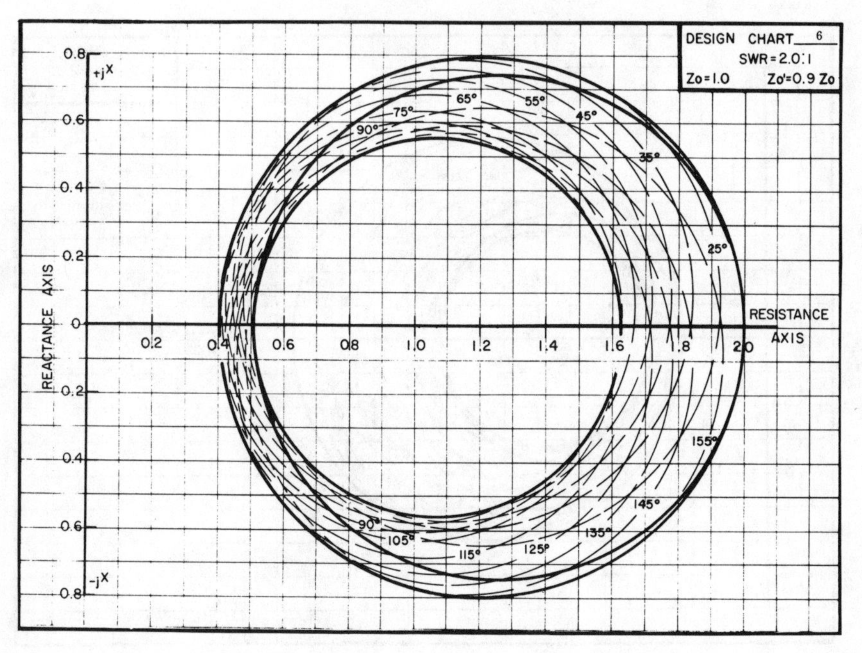
DESIGN CHART 6
SWR = 2.0:1
Zo = 1.0 Zo' = 0.9 Zo
REACTANCE AXIS
RESISTANCE AXIS
+jX
-jX
0.8
0.6
0.4
0.2
0
0.2
0.4
0.6
0.8
0.2
0.4
0.6
0.8
1.0
1.2
1.4
1.6
1.8
2.0
65°
55°
75°
45°
90°
35°
25°
155°
145°
90°
105°
135°
115°
125°

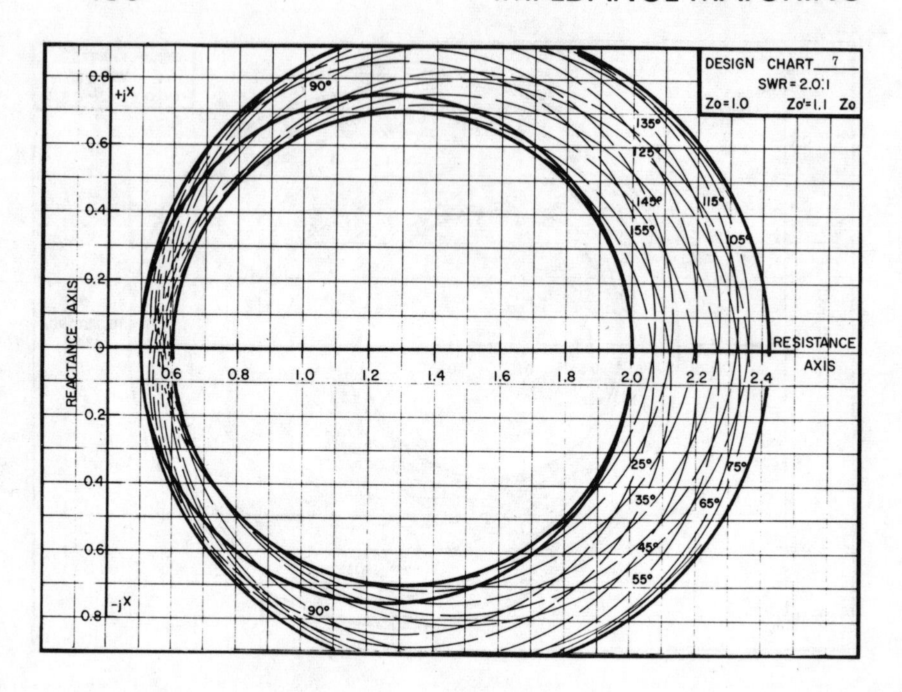
DESIGN CHART 7
SWR = 2.0:1
Zo = 1.0 Zo' = 1.1 Zo
REACTANCE AXIS
+jX
-jX
RESISTANCE AXIS
0.8
0.6
0.4
0.2
0
0.2
0.4
0.6
0.8
0.6
0.8
1.0
1.2
1.4
1.6
1.8
2.0
2.2
2.4
90°
135°
125°
145°
115°
155°
105°
25°
75°
35°
65°
45°
55°
90°

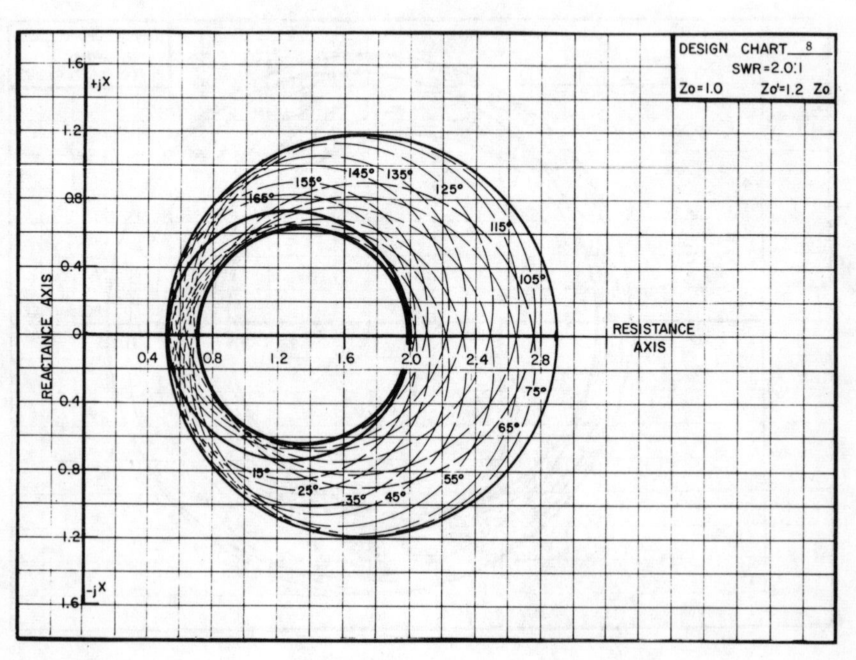
DESIGN CHART 8
SWR = 2.0:1
Zo = 1.0 Zo' = 1.2 Zo
REACTANCE AXIS
+jX
-jX
RESISTANCE AXIS
1.6
1.2
0.8
0.4
0
0.4
0.8
1.2
1.6
0.4
0.8
1.2
1.6
2.0
2.4
2.8
145°
135°
155°
125°
165°
115°
105°
75°
65°
15°
55°
25°
35°
45°

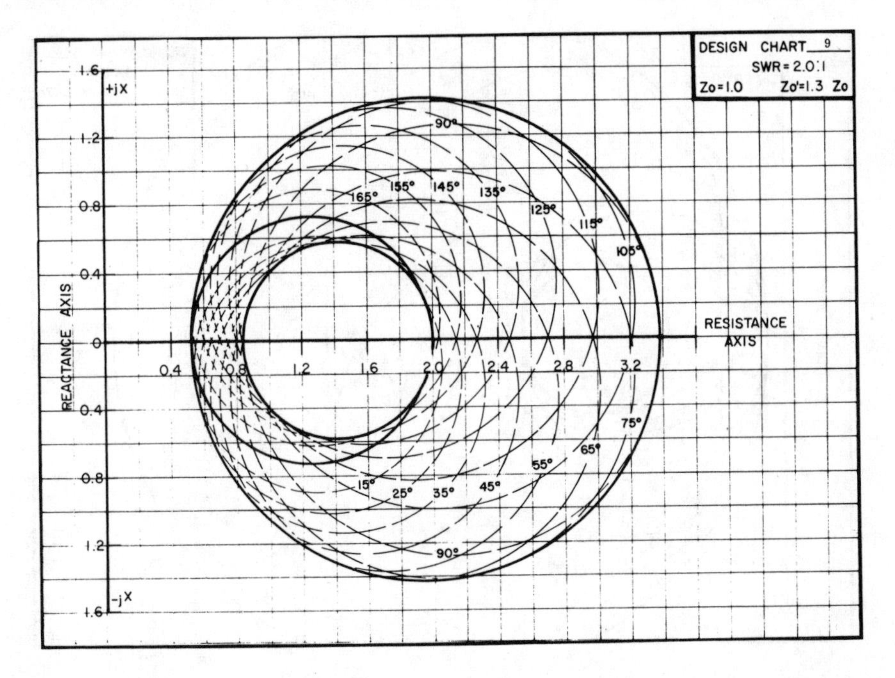
DESIGN CHART 9
SWR = 2.0:1
Zo = 1.0
Zo' = 1.3 Zo
+jX
-jX
REACTANCE AXIS
RESISTANCE AXIS
1.6
1.2
0.8
0.4
0
0.4
0.8
1.2
1.6
0.4
0.8
1.2
1.6
2.0
2.4
2.8
3.2
90°
165°
155°
145°
135°
125°
115°
105°
75°
65°
55°
15°
25°
35°
45°
90°

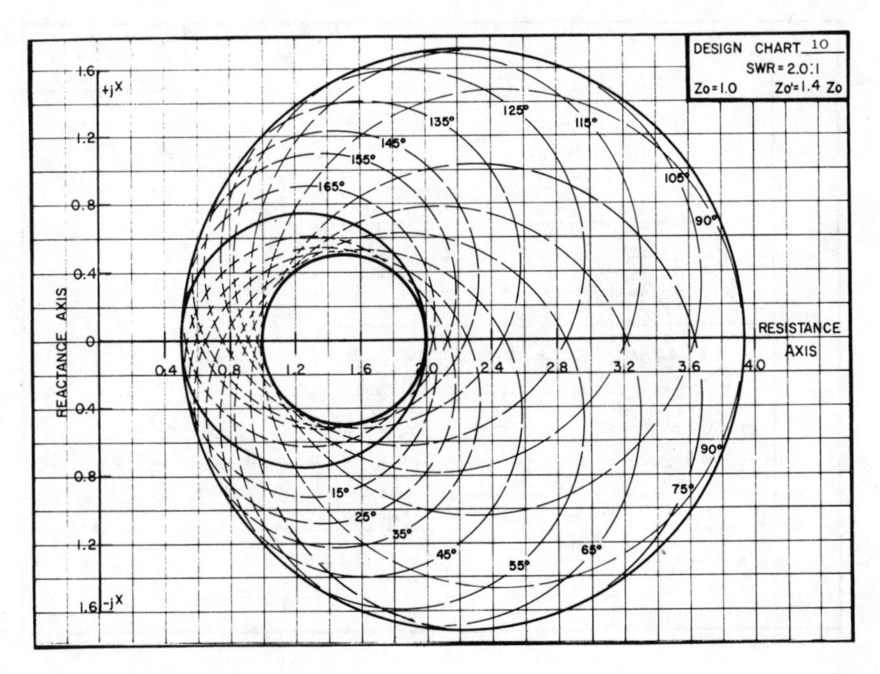
DESIGN CHART 10
SWR = 2.0:1
Zo = 1.0
Zo' = 1.4 Zo
+jX
-jX
REACTANCE AXIS
RESISTANCE AXIS
1.6
1.2
0.8
0.4
0
0.4
0.8
1.2
1.6
0.4
0.8
1.2
1.6
2.0
2.4
2.8
3.2
3.6
4.0
135°
125°
115°
145°
155°
165°
105°
90°
90°
75°
15°
25°
35°
45°
55°
65°

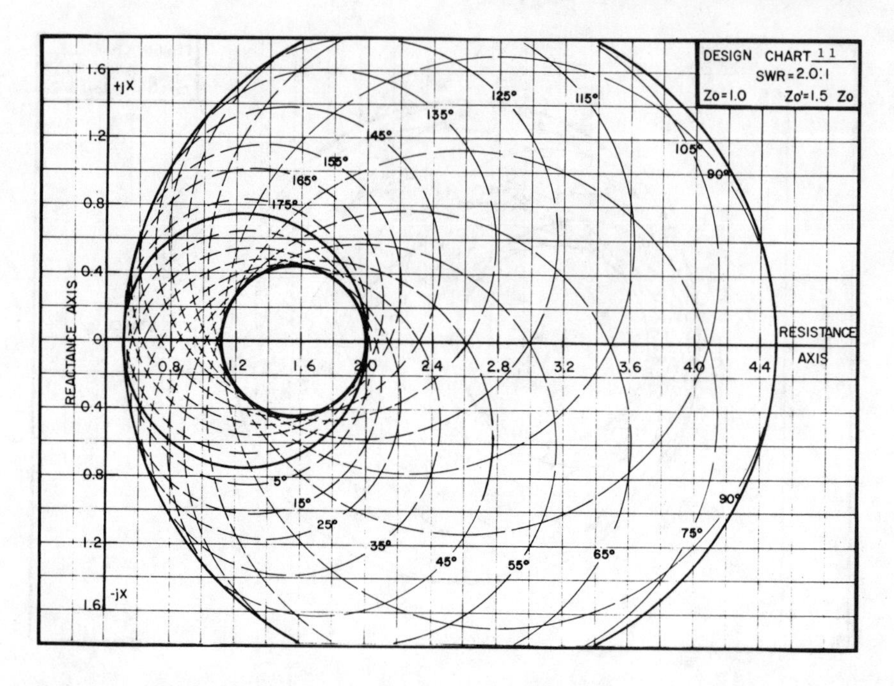
DESIGN CHART 11
SWR = 2.0:1
Zo=1.0 Zo'=1.5 Zo
+jX
-jX
1.6
1.2
0.8
0.4
0
REACTANCE AXIS
RESISTANCE AXIS
0.8
1.2
1.6
2.0
2.4
2.8
3.2
3.6
4.0
4.4
125°
115°
135°
145°
155°
165°
175°
105°
90°
5°
15°
25°
35°
45°
55°
65°
75°
90°

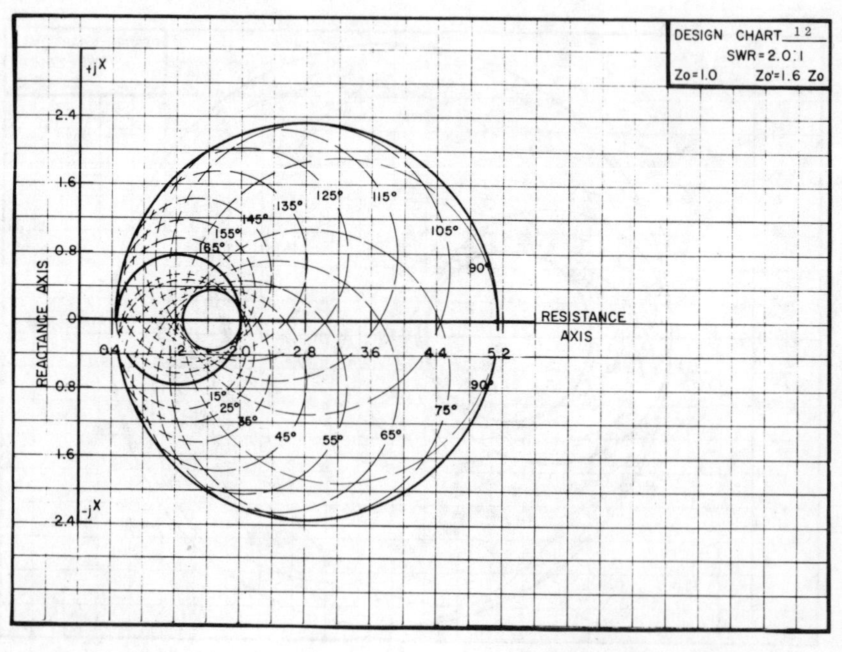
DESIGN CHART 12
SWR = 2.0:1
Zo=1.0 Zo'=1.6 Zo
+jX
-jX
2.4
1.6
0.8
0
REACTANCE AXIS
RESISTANCE AXIS
0.4
1.2
2.0
2.8
3.6
4.4
5.2
125°
115°
135°
145°
155°
165°
105°
90°
15°
25°
35°
45°
55°
65°
75°
90°

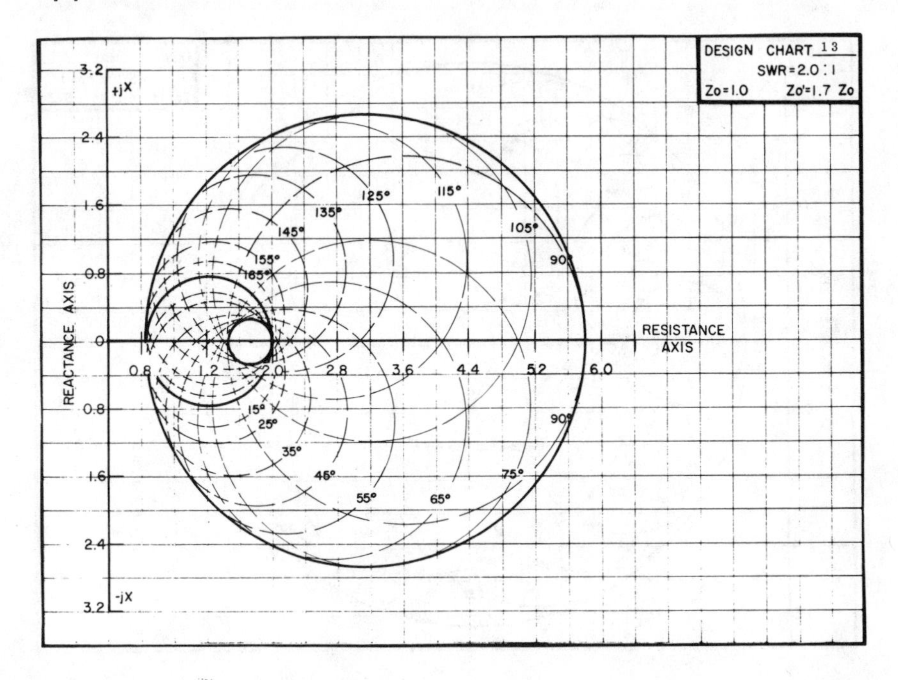
DESIGN CHART 13
SWR = 2.0 : 1
Zo = 1.0 Zo' = 1.7 Zo
REACTANCE AXIS
+jX
-jX
3.2
2.4
1.6
0.8
0
0.8
1.6
2.4
3.2
RESISTANCE AXIS
0.8
1.2
2.0
2.8
3.6
4.4
5.2
6.0
125°
115°
135°
145°
105°
155°
165°
90°
15°
25°
90°
35°
45°
75°
55°
65°

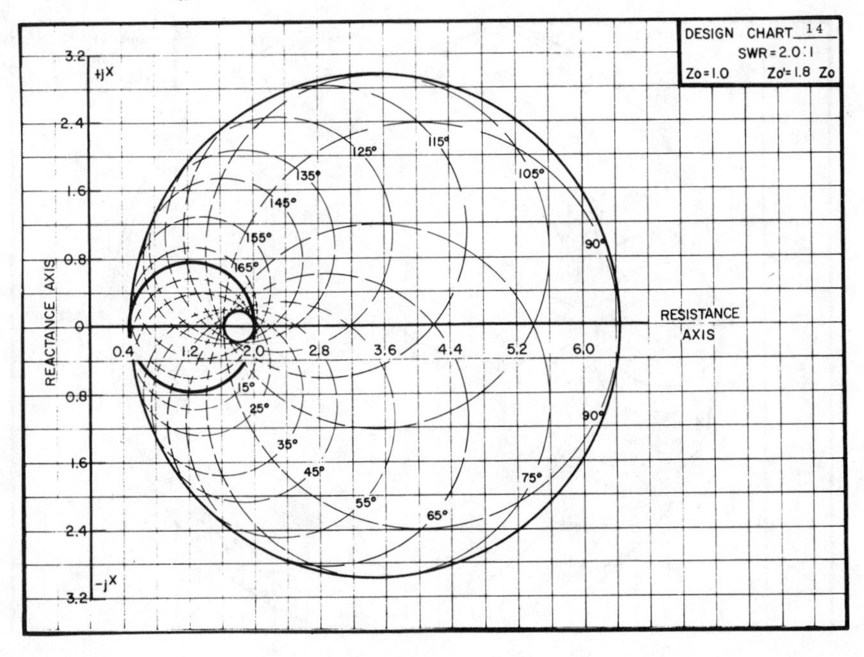
DESIGN CHART 14
SWR = 2.0 : 1
Zo = 1.0 Zo' = 1.8 Zo
REACTANCE AXIS
+jX
-jX
3.2
2.4
1.6
0.8
0
0.8
1.6
2.4
3.2
RESISTANCE AXIS
0.4
1.2
2.0
2.8
3.6
4.4
5.2
6.0
125°
115°
135°
105°
145°
155°
90°
165°
15°
25°
90°
35°
45°
75°
55°
65°

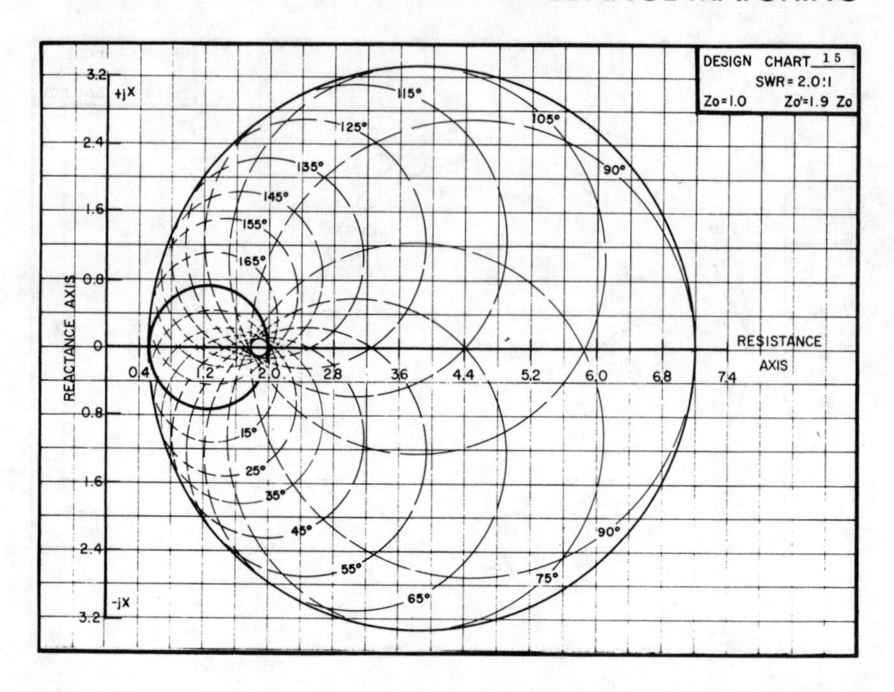

DESIGN CHART 15
SWR = 2.0:1
Zo=1.0 Zo'=1.9 Zo
+jX
-jX
REACTANCE AXIS
RESISTANCE AXIS
3.2
2.4
1.6
0.8
0
0.8
1.6
2.4
3.2
0.4
1.2
2.0
2.8
3.6
4.4
5.2
6.0
6.8
7.4
115°
125°
105°
135°
90°
145°
155°
165°
15°
25°
35°
45°
90°
55°
75°
65°

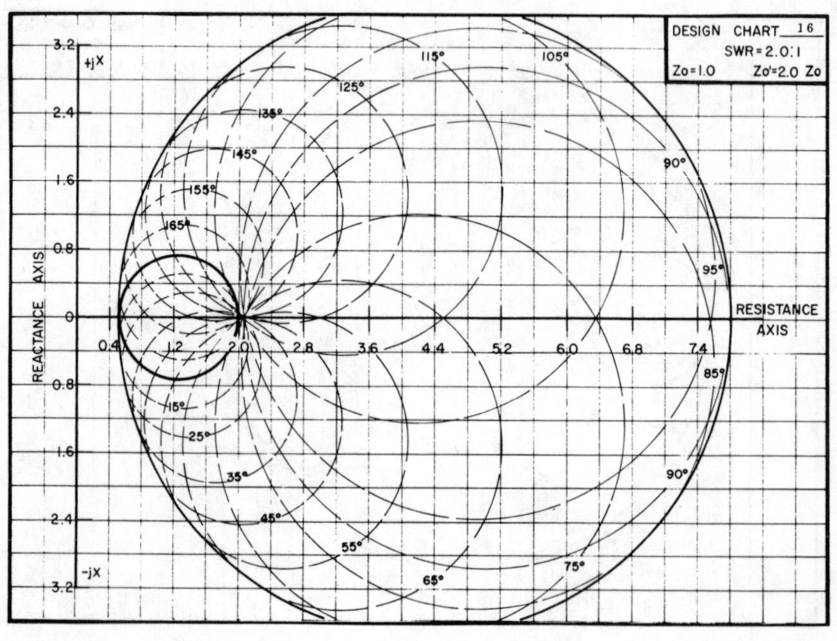

DESIGN CHART 16
SWR = 2.0:1
Zo=1.0 Zo'=2.0 Zo
+jX
-jX
REACTANCE AXIS
RESISTANCE AXIS
3.2
2.4
1.6
0.8
0
0.8
1.6
2.4
3.2
0.4
1.2
2.0
2.8
3.6
4.4
5.2
6.0
6.8
7.4
115°
105°
125°
135°
145°
90°
155°
165°
95°
85°
15°
25°
35°
90°
45°
55°
75°
65°

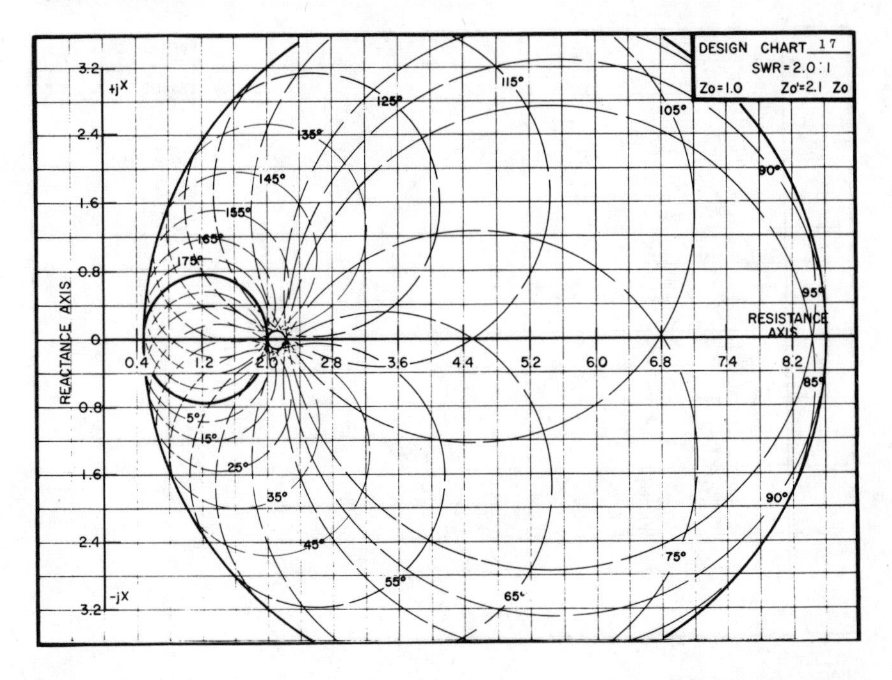
DESIGN CHART 17
SWR = 2.0:1
Zo = 1.0 Zo' = 2.1 Zo
+jX
-jX
REACTANCE AXIS
RESISTANCE AXIS
3.2
2.4
1.6
0.8
0
0.8
1.6
2.4
3.2
0.4
1.2
2.0
2.8
3.6
4.4
5.2
6.0
6.8
7.4
8.2
115°
125°
105°
135°
145°
90°
155°
165°
175°
95°
85°
5°
15°
25°
35°
90°
45°
75°
55°
65°

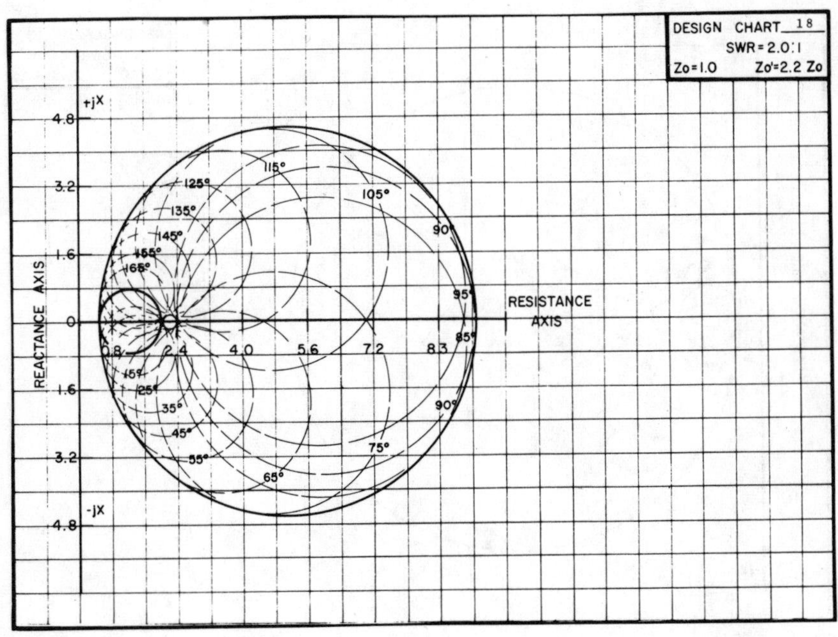
DESIGN CHART 18
SWR = 2.0:1
Zo = 1.0 Zo' = 2.2 Zo
+jX
-jX
REACTANCE AXIS
RESISTANCE AXIS
4.8
3.2
1.6
0
1.6
3.2
4.8
0.8
2.4
4.0
5.6
7.2
8.3
115°
125°
105°
135°
145°
90°
155°
165°
95°
85°
15°
25°
35°
90°
45°
75°
55°
65°

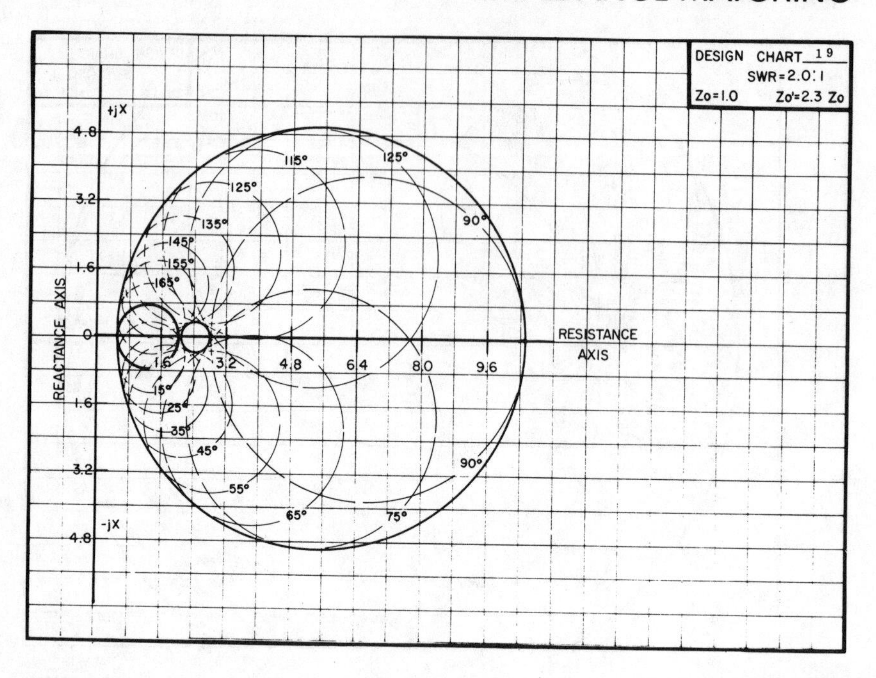
DESIGN CHART 19
SWR = 2.0:1
Zo = 1.0
Zo' = 2.3 Zo
+jX
-jX
REACTANCE AXIS
RESISTANCE AXIS
4.8
3.2
1.6
0
1.6
3.2
4.8
1.6
3.2
4.8
6.4
8.0
9.6
115°
125°
125°
135°
145°
155°
165°
90°
15°
25°
35°
45°
55°
65°
75°
90°

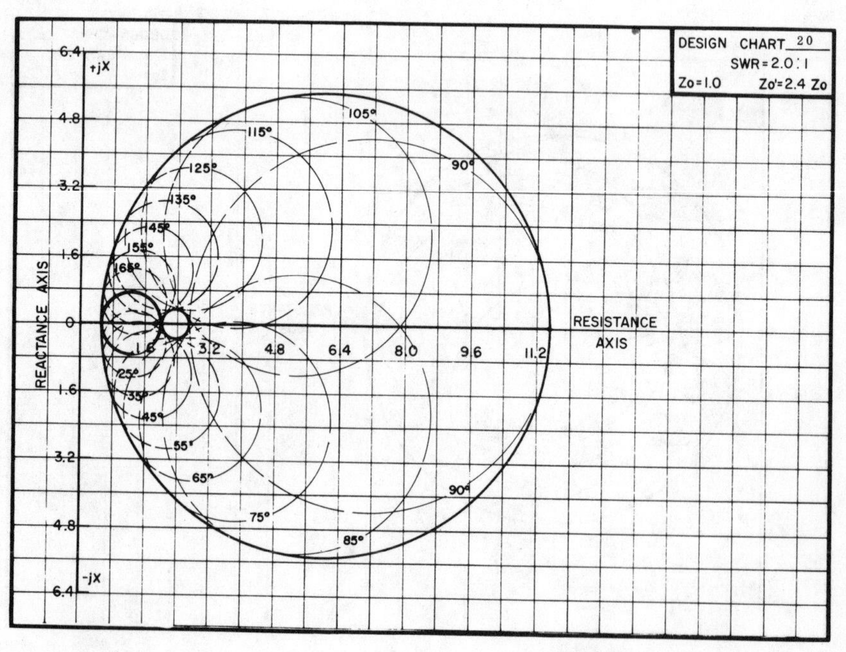
DESIGN CHART 20
SWR = 2.0:1
Zo = 1.0
Zo' = 2.4 Zo
+jX
-jX
REACTANCE AXIS
RESISTANCE AXIS
6.4
4.8
3.2
1.6
0
1.6
3.2
4.8
6.4
1.6
3.2
4.8
6.4
8.0
9.6
11.2
105°
115°
125°
135°
145°
155°
165°
90°
25°
35°
45°
55°
65°
75°
85°
90°

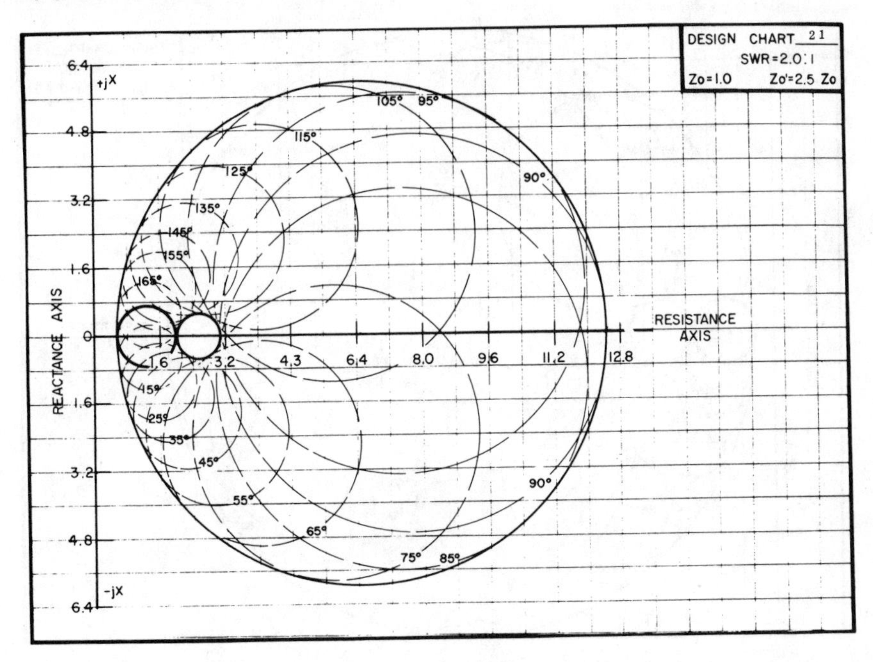
DESIGN CHART 21
SWR = 2.0:1
Zo = 1.0 Zo' = 2.5 Zo
REACTANCE AXIS
+jX
-jX
RESISTANCE AXIS
6.4
4.8
3.2
1.6
0
1.6
3.2
4.8
6.4
1.6
3.2
4.3
6.4
8.0
9.6
11.2
12.8
105°
95°
115°
125°
90°
135°
145°
155°
165°
15°
25°
35°
45°
90°
55°
65°
75°
85°

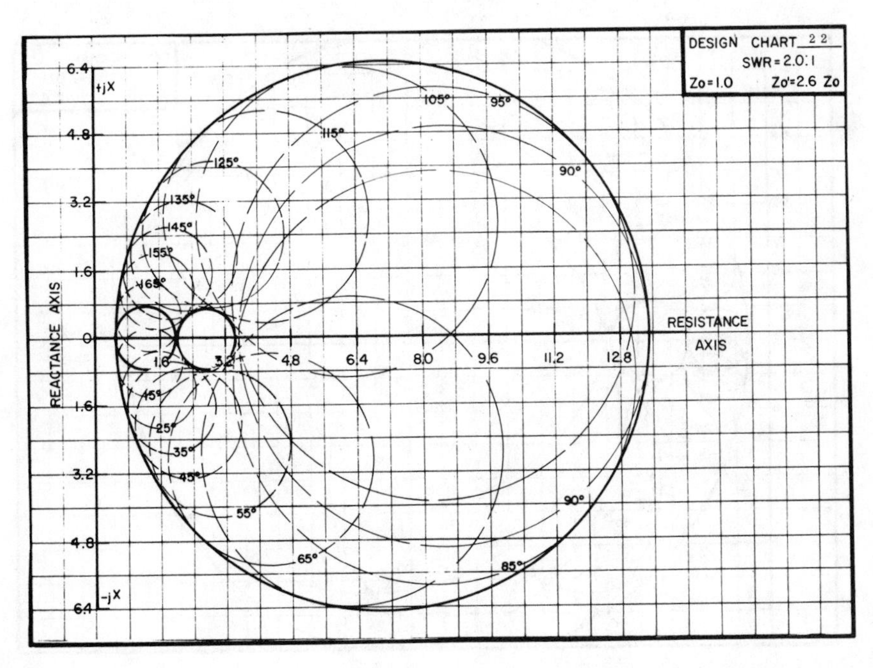
DESIGN CHART 22
SWR = 2.0:1
Zo = 1.0 Zo' = 2.6 Zo
REACTANCE AXIS
+jX
-jX
RESISTANCE AXIS
6.4
4.8
3.2
1.6
0
1.6
3.2
4.8
6.4
1.6
3.2
4.8
6.4
8.0
9.6
11.2
12.8
105°
95°
115°
125°
90°
135°
145°
155°
165°
15°
25°
35°
45°
55°
90°
65°
85°

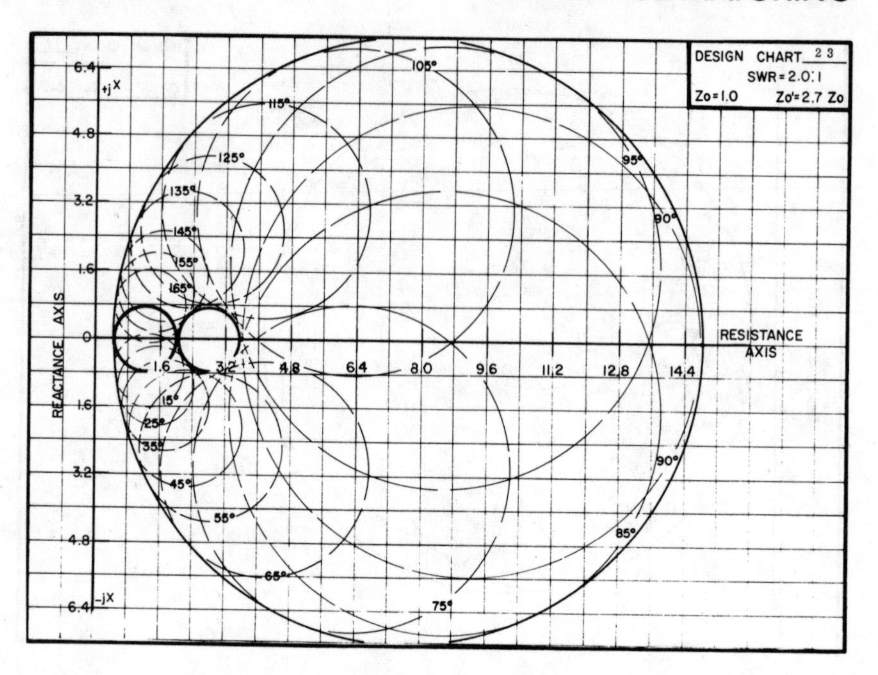
DESIGN CHART 23
SWR = 2.0:1
Zo = 1.0 Zo' = 2.7 Zo
REACTANCE AXIS
RESISTANCE AXIS
+jX
-jX
6.4
4.8
3.2
1.6
0
1.6
3.2
4.8
6.4
1.6
3.2
4.8
6.4
8.0
9.6
11.2
12.8
14.4
105°
115°
125°
135°
145°
155°
165°
95°
90°
15°
25°
35°
45°
55°
65°
75°
85°
90°

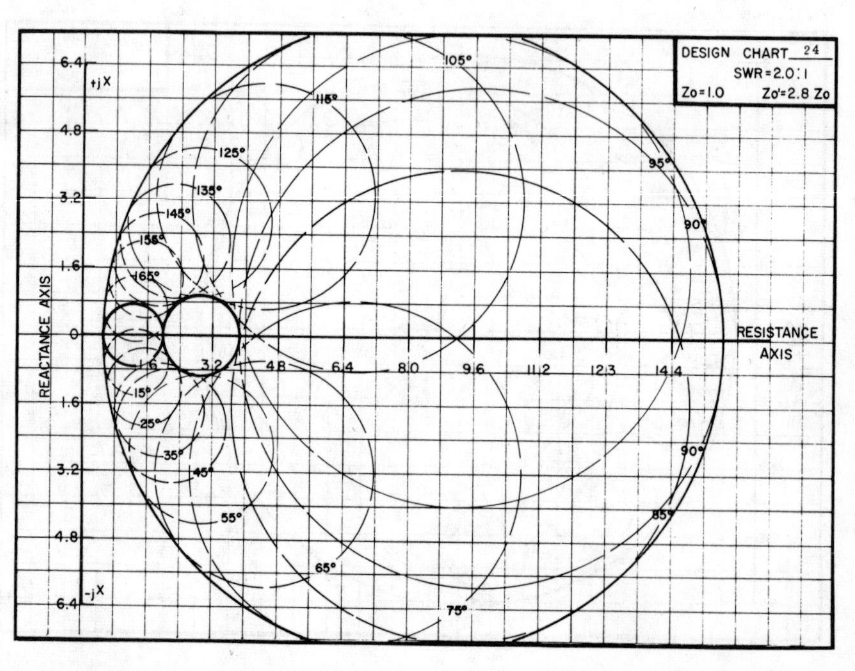
DESIGN CHART 24
SWR = 2.0:1
Zo = 1.0 Zo' = 2.8 Zo
REACTANCE AXIS
RESISTANCE AXIS
+jX
-jX
6.4
4.8
3.2
1.6
0
1.6
3.2
4.8
6.4
1.6
3.2
4.8
6.4
8.0
9.6
11.2
12.3
14.4
105°
115°
125°
135°
145°
155°
165°
95°
90°
15°
25°
35°
45°
55°
65°
75°
85°
90°

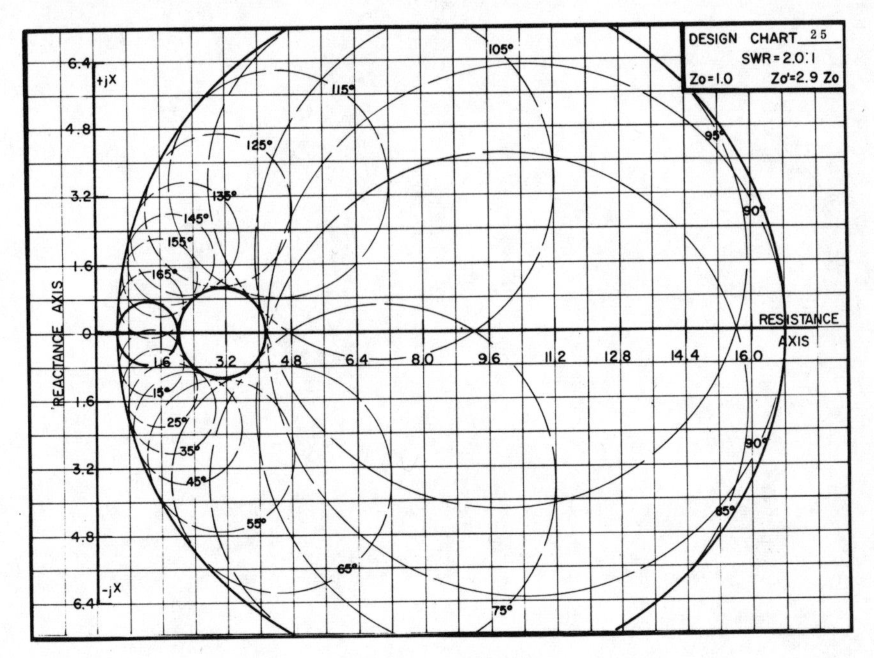
DESIGN CHART 25
SWR = 2.0:1
Zo = 1.0 Zo' = 2.9 Zo
REACTANCE AXIS
+jX
-jX
RESISTANCE AXIS
6.4
4.8
3.2
1.6
0
1.6
3.2
4.8
6.4
1.6
3.2
4.8
6.4
8.0
9.6
11.2
12.8
14.4
16.0
105°
115°
125°
135°
145°
155°
165°
95°
90°
15°
25°
35°
45°
55°
65°
75°
85°
90°

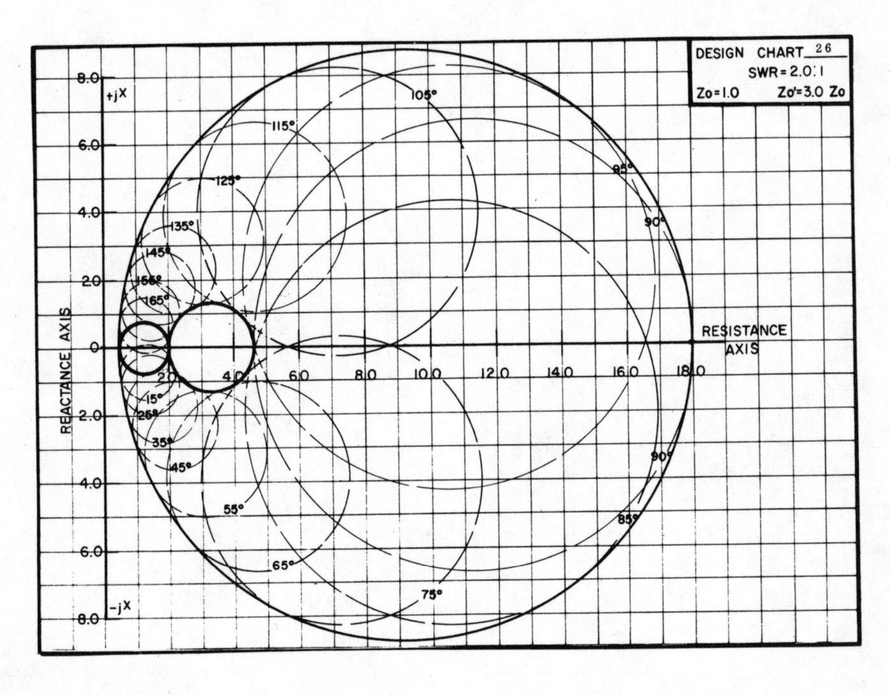
DESIGN CHART 26
SWR = 2.0:1
Zo = 1.0 Zo' = 3.0 Zo
REACTANCE AXIS
+jX
-jX
RESISTANCE AXIS
8.0
6.0
4.0
2.0
0
2.0
4.0
6.0
8.0
2.0
4.0
6.0
8.0
10.0
12.0
14.0
16.0
18.0
105°
115°
125°
135°
145°
155°
165°
95°
90°
15°
25°
35°
45°
55°
65°
75°
85°
90°